AF866373

STOP CHASING AGILE

FROM PROCESS WORSHIP TO REAL PROJECT PERFORMANCE IN A CONSTANTLY CHANGING WORLD

JEREMY BERRIAULT

J.ROSS PUBLISHING

ISBN: 978-1-60427-217-8

e-ISBN: 978-1-60427-872-9

Printed and bound in the U.S.A. Printed on acid-free paper.

10 9 8 7 6 5 4 3 2 1

Library of Congress Cataloging-in-Publication Data can be found in theWAV section of the publisher's website at www.jrosspub.com/wav.

Direct all inquiries to J. Ross Publishing, Inc., 151 N. Nob Hill Rd., Suite 476, Plantation, FL 33324. For EU safety/GPSR concerns, email: gpsr@mare-nostrum.co.uk or contact: Mare Nostrum Group B.V, Doelen 72, 4831 GR Breda, The Netherlands.

Phone: (954) 727-9333
Fax: (561) 892-0700
Web: www.jrosspub.com

To my father.

And to my uncles, Brian and Jerry. You helped shape how I think about work, craft, and the value of the right tool at the right time. I miss you both.

To my father

And to my uncles, [illegible] and Jerry. [illegible] I think [illegible] work [illegible] the [illegible]
[illegible]

CONTENTS

ACKNOWLEDGMENTS

Writing a book is weird. You sit alone, argue with yourself, and try to turn years of work into sentences that don't suck. You don't get through that without people around you keeping you honest.

To the clients and teams who let me into their world: thank you. You trusted me with the real stuff: conflict, pressure, politics, burnout, and the occasional miracle. This book is built from that reality.

To the people who reviewed drafts and gave feedback: thank you for calling out the weak parts, the repetitive parts, and the moments where I was being a bit of an ass. You were right more often than I wanted to admit. And you made the book better.

To the great leaders and coaches I've worked with: thank you for showing me what service looks like when it's not a LinkedIn post. You modeled the kind of leadership that actually helps people. And to the not-so-great ones: thanks for the material.

I also want to acknowledge the thinkers who shaped our understanding of learning, systems, leadership, and teamwork long before "agile influencer" became a job title. The ones who did the work shared what they learned and didn't try to turn it into a religion. Even when I disagree with parts of their thinking, I respect the contribution.

This book isn't a rejection of learning. It's a rejection of the grift.

To my family and friends: thank you for your patience while I disappeared into writing. I owe you time back.

And to anyone who's been going through the motions and thought, *There has to be a better way*: this book is for you.

ACKNOWLEDGMENTS

INTRODUCTION

Why am I saying, "Stop Chasing Agile?" Why now? And why am I the one saying it to you?

A few years ago, I wrote *Beyond The Framework: Cultivating Agile Growth*. It was a guide I built and refined over time, based on what I saw actually work in organizations trying to get better, not just look better. The point wasn't to pick the right framework. The point was building the conditions where improvement can stick.

In business, being "agile" isn't a badge. It's the ability to respond fast when the market shifts, customers change their minds, or competitors move. That takes more than a process. It takes an *ethos* and a culture that support adaptability, creativity, and collaboration—especially when things get tense. And here's what I learned after that book: even when people understand this, they still get pulled back into the worshiping process. They start chasing labels. They start copying ceremonies. They start measuring activity instead of outcomes.

That's where this book comes in. Over the past decade, I've heard the same tired lines on repeat:

"Agile is dead."

"Agile doesn't work here."

"We're too big to be Agile."

"We're too small to be Agile."

Excuse after excuse after excuse. Most of it comes from the same sources: bad experiences, misinformation, and weak leadership. Honestly, it drives me nuts. I'm tired of people saying Agile doesn't work when what they really mean is: We did it badly and don't want to admit it.

Years ago, Dave Thomas said something that should've ended this whole debate: "Agile" is an adjective. When the *Manifesto for Agile Software Development* was written, that's how the word was meant to be used. Go read it. The word "Agile" describes *how* we work. It's not a brand. It's not a framework. It's not a certification ladder. In his GOTO talk, *Agile Is Dead (Long Live Agility)*, he makes the same point I'm making here: agility is about how you behave and adapt—not what you buy, and not what you call yourself.

But somewhere along the way, we messed it up. What started as a push for real change got buried under layers of nonsense. The industry turned into a disorganized mess. "Agile" became a label you can slap on anything: a process, a tool, a job title, a consulting package. "We're Agile now" became a slogan instead of a real shift in how people behave, decide, and lead. So when people say, "Agile is dead," I get it. If *that's* what you think Agile is, maybe it should be. But if your version of Agile is dead, that's not a funeral I'm attending. You can't kill an adjective. You can't kill Agile sales, Agile leadership, Agile development, or Agile marketing. Or anything else you want to make more adaptive.

Agile describes how something works. At its core, agility is about being adaptive, getting better than we were before by cutting red tape, becoming more self-sufficient, moving away from top-down command-and-control, and practicing adaptive leadership. It's not a badge. It's not a personality. It's a response to reality. And reality doesn't move at 1950s speed anymore. This is the 21st century. Information doesn't trickle out slowly and stay put for decades. Customer expectations shift. Markets move. Competitors copy what works. Regulations change. Technology changes. Sometimes the ground moves under your feet in a single quarter. In that environment, the old way isn't just cozy, it's costly.

So, when someone says, "Only this can work," or "We can't be Agile because X," what they're often really saying is: *I'm attached to the old way, and I don't want to change.* That's one of the most annoying things to hear, because it's giving up before you even try. It's also weird because everywhere else in life, we preach persistence. Don't give up. Try again. Keep pushing. But when it comes to agility, suddenly it's, "Pfft. Let's not try. It's new. The old way is nice and cozy."

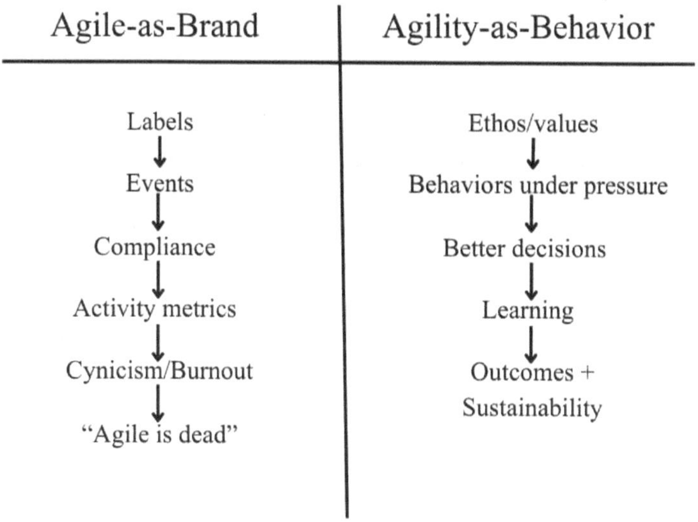

Figure I.1 | Same word. Two very different realities

Most of the ideas people now call Agile leadership, servant leadership, or transformational leadership have been around for decades. This isn't new stuff. We're just finally admitting maybe we should use it. And yet, instead of moving forward, we get stuck in the same arguments on repeat: "Agile is dead." "Leadership will never buy in." "Our culture won't change." I've lost count of how many times I've heard people say, "That won't work here," like it's a law of physics instead of a choice (see Figure I.1).

That's where the title of this book comes from. Stopping the chase isn't about burning it all down and going back to Waterfall. It's about stepping out of the noise and getting back to what matters: being better at what we do, using the tools that help, and dropping the ones that don't. Scrum, SAFe™, LeSS, Kanban, whatever you use, those are tools. Useful sometimes. Harmful sometimes. Never holy.

The point is simpler than the industry wants it to be: are we delivering value, learning, improving, and are people burning out or thriving? Right now, too many organizations are stuck. They'd rather argue about frameworks than fix behavior.

Ethos Before Mindset

One of the biggest blockers to real agility isn't process. It's a mix of cognitive bias, logical fallacies, and ego, especially in leadership. There's still this belief that if you're a manager, VP, or executive, you're supposed to have all the answers. You're the boss. You're the smartest one in the room. You decide; everyone else executes. That's not leadership. That's outdated, ego-driven nonsense.

Organizations are built on collective intelligence. If you've got more than one person in your company, the group is more intelligent than any single individual. Full stop. The moment a leader defaults to "what I say goes," they shrink the organization to the size of their own ego. Real leaders do the opposite: they focus on improving everyone around them. They build people up and create an environment where teams can think, experiment, and grow, without fear.

I learned this firsthand as a senior executive. Early in my leadership career, it took me a while to realize that my top priority wasn't running the show. It was helping the people around me progress. Their growth was the only way anything would ever work efficiently and effectively for the long haul. When leaders treat people like replaceable parts, they get compliance. When leaders treat people like capable adults, they get judgment, ownership, and adaptation, the stuff everyone claims they want. That's why I talk about *ethos* (see Figure I.2).

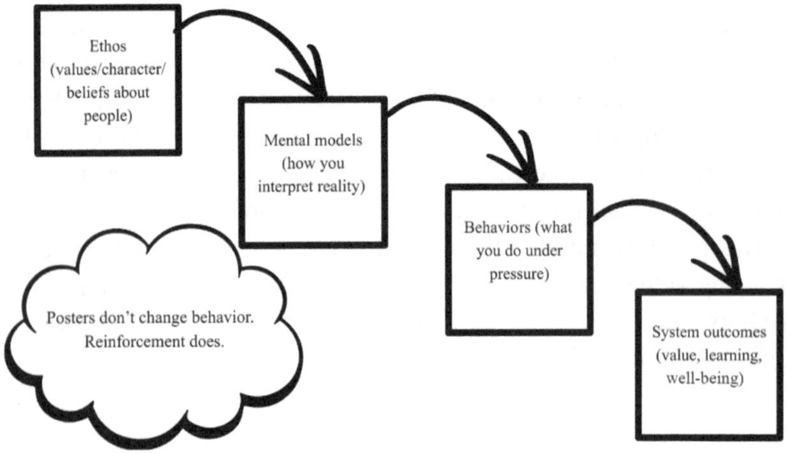

Figure I.2 | Posters don't change behavior, ethos does

Ethos shapes mental models, and mental models shape what people call *mindset*. The industry loves the phrase "the Agile mindset." Honestly? Hard pass. You don't start with a mindset. You start with ethos: your character, your values, and what you believe about people and work. Ethos isn't set in stone; we can grow, change, and rebuild our character over time (I'm one of those stories). But it's still the foundation. If you don't examine it, any mindset change turns into window dressing.

You can go to all the Agile trainings you want, memorize the values and principles, and still behave like a command-and-control dictator on Monday morning. Real change starts with what you truly believe, not what you put on a poster. That's a big reason so many Agile transformations have failed over the last 20+ years: they never changed character or values. They forced teams to "do Scrum" or "do SAFe™" and hoped speed would magically lead to success.

Sometimes things did move faster. But if you're speeding up bad decisions or poor quality, you're not improving; you're just producing bad results more quickly. Garbage in, garbage out… only now it happens at lightning speed. Here's a real example of how this plays out. I once worked on a massive system rewrite that replaced an entire legacy platform. The stated goal wasn't better decisions, better data quality, or better customer outcomes. It was simply: Get the data faster. They were receiving data after about two weeks and wanted it within one. That was the win they were chasing. The part that still blows my mind is that they already knew the data was wrong. They weren't fixing the quality or the underlying issues. They just wanted bad data delivered faster so people could clean it up faster. Think about that for a second. They went through all the motions—sprints, iterations, chunked work, ceremonies. On paper, it looked "Agile." In practice, it was a costly way to do the wrong thing more quickly.

The program was planned for two years. After I left unceremoniously for raising uncomfortable but valid points, it ended up taking about four times as long as planned, and the budget increased substantially. Did they eventually get something that worked? Maybe. I spoke with the person who replaced me later, and he couldn't even find the words to describe how it went. The whole thing was a perfect example of what happens when you prioritize speed and process over character, values, and outcomes.

Why This Book, Really

This book is my attempt to shut out the noise and get us back to what matters. Agility works, and it's needed now more than ever, whether you're building software, running a mom-and-pop shop, operating a retail store, leading a marketing firm, or working in the service industry. Real agility (not the branded, buzzword version) is what helps you survive change, make better decisions under pressure, and keep improving without burning people out.

If you think you can keep managing with old-school, top-down, I-have-all-the-answers leadership and still innovate, you're lying to yourself. This is a group effort. Organizations don't get better because one smart person has a strong opinion. They get better when leaders create the conditions for collective intelligence to emerge: clear priorities, quick decisions, honest feedback, and a safe space to experiment. This book is here to help you spot what's really getting in the way, understand why so many Agile efforts fail, and shift from mindset talk to ethos and behavior. My goal is simple: help you build something that works. Not perfect. Not trendy. Not branded. Just better.

With a title like *Stop Chasing Agile*, you might be wondering where this is going. Maybe you expected a framework takedown. Maybe you're waiting for the how-to part. Fair. Let's clear it up right now.

This book isn't a framework manual. I'm not here to teach you how to do Scrum properly, master SAFe™, or chase the latest shiny thing (whatever's trending this quarter). Frameworks are tools. They can be useful. They can also become a distraction. Either way, they're not the point. If what you want is step-by-step instructions, then YouTube, online courses, and certifications have you covered.

It's also not a "10 Steps to Agile" guide. I'm not serving up "7 Secrets to Better Retros," "12 Hacks to Transform Your Stand-Ups," or a checklist you can follow without thinking. There are already plenty of books and blogs for that. Some of them are even decent. But most of it still assumes the same thing: if you install the right process, behavior will magically follow. This book is about what actually makes agility work—regardless of the framework you use (or don't). It's about the human factors: ethos, decision-making, conflict, clarity, feedback, and what leaders do when the plan gets punched in the face. That's the leverage. That's the part most Agile books politely step around. So no, you won't find a paint-by-numbers guide here. But if you're ready to stop worshiping process and start leading like a human, let's get to work.

So What Is This Book?

It's a noise filter. A practical lens for seeing what agility is, what gets in the way, and how to tell the difference between something that will actually help you and something that's just branding, ego, or recycled advice dressed up as something new. Because let's be honest: there's a lot of crap out there, and there's good thinking too. The problem is that most people don't have a clear way to separate signal from noise when they're being hit with frameworks, labels, certifications, and thought leaders all promising that this is the thing that will finally make Agile work. This book is here to help you cut through that, spot what's useful for you and your team, and understand agility as an environment, *an ecosystem,* not just a process. I'm not here to sell you my own framework. I'm here to help you see clearly and make better choices for your context.

This book is for business owners and founders, senior leaders and executives, team leaders and managers, coaches and Agile practitioners, and team members who want work to suck less and matter more. It's especially for people who've been burned by bad Agile: the big "transformation" that started with a rah-rah kickoff and died six months later; the rollout that created more frustration, more bureaucracy, and less trust; the experience that left you thinking, *If this is Agile, I don't want it.* If you've been through that and you're still here, still reading, it's probably because you believe, somewhere under the frustration, that work can be better. That teams can be healthier. That agility should mean something more than "we do stand-ups now." If that's you, you're in the right place.

This book isn't for people looking for a new label to slap on what they already do. It isn't for badge hunters chasing the next certification. And it definitely isn't for anyone who just wants their favorite framework validated and defended. If someone tries to build a certification out of this book, they've missed the point entirely.

I wrote this to open eyes, not to create another club. I want you to see what real agility means for your organization, your team, your leaders, and for you as an individual contributor. This is about becoming better, not more "on brand," not more compliant. Just better. And yes, this book comes from frustration with the current state of the Agile world. But it also comes from hope: hope that we can stop the worshiping process, stop arguing in circles, and move forward.

How This Book Is Structured

This book weaves together four main threads. You'll see each one show up across the chapters, stories, and tools, sometimes quietly in the background, sometimes right in your face.

First, I name the noise. Not to be edgy for sport, but because a lot of organizations are stuck in patterns they can't describe. When you can't name what's happening, you can't change it. Second, I draw on decades of research and thinking about systems, leadership, learning, and human behavior that predated Agile Influencer becoming a job title. Third, the book stays behavior-first and ethos-first. I'm not interested in the installation process and hoping behavior follows. I'm interested in what people do under pressure, what leaders reward, and what teams learn to tolerate. Finally, I ground it in real stories, case studies, and small changes that actually work in the mess of real organizations. Let's break those down.

Calling Out the Nonsense

You'll see this thread in chapters like The Agile Graveyard, The Framework Fiasco, and the sections that follow influencer culture and what Dean Peters called *hyper-scrum-mentalism*, the habit of treating one framework as gospel and everything else as heresy.[1] If you've spent any time on LinkedIn, you've seen the graveyard posts: "This is dead." "That is dead." "Here's the new phoenix; this is the optimal framework now." It's ridiculous. Most of it is rebranding, status games, and ego. This section is about naming what's happening: framework worship,

1 Dean Peters, "The 4 'isms' of Agile Failures," *Medium*, 2018, https://medium.com/deanondelivery/the-4-isms-of-agile-failures-a40c960ace1f. Peters' idea of "hyper-scrum-mentalism" and his broader critique of Agile dogma were one of the sparks that pushed me toward writing this book.

certification culture, influencer economics, and shiny new labels slapped on the same old dysfunction. Not because I love ranting (okay, sometimes I do), but because you can't fix what you won't name.

Bringing in the Research (That's Been Around for Decades)

Most of what we'll talk about in this book isn't new. Research on motivation, team dynamics, leadership, and systems thinking has long existed, often dating back to the last century and, in some cases, even earlier.[2] Agile isn't some brand-new invention. The language is newer, sure, but the principles underneath it are old and well tested. People do better work when they have autonomy, mastery, and purpose. Teams outperform lone heroes when properly set up. Systems shape behavior far more than slogans do. Organizations that learn faster tend to stick around longer. None of that is controversial in the research. What's controversial is how often we ignore it in practice, then act surprised when a transformation turns into frustration. This part brings that research into the book for one reason: not to turn this into an academic slog, but to give you something solid to stand on. If you've been feeling like modern work has drifted into nonsense—more meetings, more process, less progress—you're not being dramatic. There's evidence behind what so many people feel in their gut. The clues have been there for decades. We've just been busy chasing new labels instead of applying what we already know.

Behavior-First and Ethos-First Agility

This thread puts behavior and ethos, your character and values, center stage. If you're wondering, "Is it behavior-first or ethos-first?" the honest answer is both. They're tied together. Your ethos shapes your behavior: the choices you make, what you avoid, what you tolerate, and how you

2 For example, on motivation and human needs, see Douglas McGregor, *The Human Side of Enterprise* (McGraw Hill, 1960); Frederick Herzberg, *Work and the Nature of Man* (World Publishing, 1966); and Edward L. Deci and Richard M. Ryan, *Intrinsic Motivation and Self-Determination in Human Behavior* (Plenum, 1985). On team dynamics and group effectiveness, see J. Richard Hackman, *Leading Teams: Setting the Stage for Great Performances* (Harvard Business School Press, 2002). On systems thinking and learning organizations, see Peter M. Senge, *The Fifth Discipline: The Art and Practice of the Learning Organization* (Doubleday, 1990).

treat people under pressure. At the same time, consistently changing behaviors can reshape your ethos. Not overnight. Not with a slogan. But through practice and reinforcement. Think of ethos as the foundation and behavior as the building blocks. Without a solid foundation, the building won't stand. But without the blocks, the foundation never becomes a home. That's why this book treats Agile values as lived practices, not words on a poster or a slide deck.[3] If your values only show up in a kickoff deck, they're not values. They're marketing. And yes, this means changing some behaviors. Maybe even confronting parts of your character. That's the uncomfortable part. But if you're still reading, I'm betting you're not here just for a better process. You want to show up better, help your team get better, and make your organization better. This isn't the "think Agile thoughts" section. This is the "look in the mirror" section.

We're going to dig into what you do under pressure, what you really believe about people and work, and how those beliefs leak into the way you lead, decide, and respond. Not "memorize the Manifesto." Not "say the right words." More like: change how you behave, change what you tolerate, and change what you reward. Until those three things shift, nothing else—frameworks, tools, events—will matter for very long.

Real Stories, Case Studies, and Small Changes

You'll get real stories in this book, some that make me look good, and some that don't. That's the point. If all you ever hear are polished success stories, you don't learn much. The valuable stuff lives in the messy middle: the times we got it wrong, the times ego got in the way, and the times we thought we were doing Agile when we were really just rearranging the furniture. Most of what works isn't dramatic. It's small, consistent changes in how we lead, communicate, decide, and learn. Those changes don't make for flashy LinkedIn posts, but over time they're what make an organization genuinely adaptive. They're also the only kind of change

3 On underlying values and assumptions shaping behavior and culture, see Edgar H. Schein, *Organizational Culture and Leadership* (Jossey-Bass, multiple editions). On mental models as the often-invisible lenses through which we interpret reality, see Peter M. Senge, *The Fifth Discipline* (Doubleday, 1990). On mindset as beliefs about learning, ability, and challenge, see Carol S. Dweck, *Mindset: The New Psychology of Success* (Random House, 2006).

that tends to survive the next re-org, the next executive shuffle, or the next new operating model announcement.

I've had to unlearn this myself. Early on, I thought leadership meant control: be the boss, have the answers, keep a tight grip. Turns out that grip was the bottleneck. Things improved when I stopped trying to control people and started fixing the system where they were trapped. Then, I helped them grow instead of squeezing them for output. That's what this book is about. Not a new framework or label, but a different way of leading real humans through hard work.

One Last Thing Before We Dive In

Before we get into the chapters, here's the deal. This book isn't here to defend a framework or hand you a new method to install. It's here to look at what actually drives results when things get messy, and what we each reinforce in the system when we're under pressure. If you're open to questioning assumptions, habits, and yes, ego, keep reading. If you're willing to look in the mirror and ask, "What am I contributing to the way work feels and flows here?" you're exactly who I wrote this for. And if what you really want is a story where the method is always right and the people are always the problem, parts of this will feel uncomfortable. That discomfort isn't the point by itself, but it can be the doorway to better conversations, clearer decisions, and a healthier way of doing hard work together.

HOW TO READ THIS BOOK (DEPENDING ON YOUR ROLE)

Not everyone comes to this book from the same place. Some of you are executives. Some are managers. Some are coaches, Scrum Masters, or consultants. Some of you are on the front lines, trying to get work done without losing your mind. You'll notice different things in here depending on where you sit, and that's a good thing.

If you're an executive or senior leader, read this book as a mirror, not as a manual. Pay close attention to the sections on ethos, power, and ego, and to the stories where leadership behavior quietly kills adaptability. Don't just focus on what teams should do. Watch for the system signals you're sending, what gets rewarded, what gets punished, and what people learn is unsafe to say out loud. As you read, ask yourself: Where am I part of the problem? What am I rewarding, real improvement or good-looking activity? What am I tolerating that I'd never admit publicly? If you only use this book to judge your teams and never question your own behavior, you'll miss the point.

If you're a middle manager or team lead, you're probably stuck in the squeeze: pressure from above, pain from below. Read this book to name what you're seeing, find language to push back constructively, and spot small, fundamental changes you can make within your span of control. You may not be able to change the whole organization, but you can change how it feels to work for you, and that matters more than most people realize. Ask yourself: What can I change in how I lead, even if nothing above me changes yet? Where am I unintentionally passing pressure and fear down the chain? What's one behavior I can model that makes it safer for my team to tell the truth?

If you're a Scrum Master, Agile coach, or consultant, this book might poke at your identity a bit. That's intentional. Read it as a challenge to any framework loyalty you've picked up over the years, and as a reminder that your job is to change behaviors and systems, not just run events. Use it as a toolkit for spotting hype, both within the industry and within your own practice. Ask yourself: Where have I been selling process instead of helping people think? Am I more attached to my favorite approach than to my client's reality? What would it look like to be behavior-first, not framework-first, in my work? If you feel defensive at points, don't rush past it. That's usually where the useful learning is.

If you're a team member, individual contributor, or frontline worker, you might not have "leader" in your title, but you're the one living with the consequences of all this. Read this book as validation that you're not crazy, as language for what you're experiencing, and as a way to spot when you're being sold something that looks good but doesn't help. Start with what's in your control right now. Look for small ways to nudge your and others' behavior in a better direction. Ask questions that challenge assumptions and invite honesty. Speak up constructively when it's safe, and when it isn't, find allies who can help carry the message. You don't have to wait for a formal transformation to start acting with more clarity, courage, and care. Ask yourself: What do I want to start asking my leaders and coaches that I haven't asked before?

If you are in hybrid roles (manager and contributor, coach and internal leader, founder and everything else), read the book once with one hat on, then again with another. Different parts will sting. Different parts will feel like relief. Different parts will give you ideas you didn't see the first time. However you read it: if you're willing to question your own assumptions, you'll get something out of it. If you're only here to defend your favorite framework, you won't. Use this book as a mirror and a starting point for better conversations, not as a new set of rules.

Happy reading!

ABOUT THE AUTHOR

Jeremy Berriault is a human-centered performance architect and founder of Berriault and Associates Consulting Group, where he helps leaders recognize when performance comes at a human cost—and rebuild teams that can adapt, perform, and sustain success. His work is grounded in a behavior-first philosophy: real agility isn't a title or framework, but how people show up and work together under pressure. Drawing on decades of experience, he has developed practical tools and operating rhythms that cut through competing priorities, sharpen decision-making, and embed change that actually lasts. Jeremy has supported leaders across industries—from software and finance to hospitality and fitness—earning recognition for helping teams build clarity, trust, and healthier execution.

A seasoned advisor, speaker, and coach, Jeremy is known for translating complex organizational challenges into clear, actionable insights. His approach blends research-backed assessments with hands-on workshops and coaching, ensuring that transformation efforts are both practical and durable. Over a career spanning more than 30 years, he has held roles from quality assurance leadership to enterprise delivery and executive positions, guiding cross-functional teams toward meaningful business outcomes. He holds an MBA and graduate management credentials from Athabasca University, along with a Bachelor of Applied Science in Psychology from York University. Outside of his professional work, he trains in jiu-jitsu—a discipline that reflects his belief in honest pressure, immediate feedback, and continuous improvement.

WAV Web Added Value™

This book has free material available for download from the Web Added Value™ resource center at *www.jrosspub.com*

At J. Ross Publishing we are committed to providing today's professional with practical, hands-on tools that enhance the learning experience and give readers an opportunity to apply what they have learned. That is why we offer free ancillary materials available for download on this book and all participating Web Added Value™ publications. These online resources may include interactive versions of material that appears in the book or supplemental templates, worksheets, models, plans, case studies, proposals, spreadsheets and assessment tools, among other things. Whenever you see the WAV™ symbol in any of our publications, it means bonus materials accompany the book and are available from the Web Added Value Download Resource Center at www.jrosspub.com.

Downloads for *Stop Chasing Agile* include the Appendices at the end of this book for use in your organization.

CHAPTER 1

THE AGILE GRAVEYARD

It's a dark, gloomy night. You're walking through an old graveyard. The air is cold enough that you can see your breath. Leaves rustle in the wind. A thin mist curls around your feet with every step, like the ground itself is exhaling. You sweep your flashlight across the grass, and the shapes start to appear: gravestones, row after row, stretching out into the fog. You step closer to one and angle the light. The engraving comes into focus:

"Scrum—Dead."

You move to the next stone:

"SAFe™—Dead."

Then another:

"Agile—Dead."

And another:

"Jira—Dead."

Everywhere you look, it's the same story. Dead. You feel the weight of all this loss pressing down on you. Everything you've worked with, everything you've tried to use to create value, is apparently buried here. Whatever hope you had starts to feel a little shaky. If all of this is dead, what's left?

What's left is what doesn't fit on a slide deck: how you show up when things get hard. Because if you're going to do any of this, you might

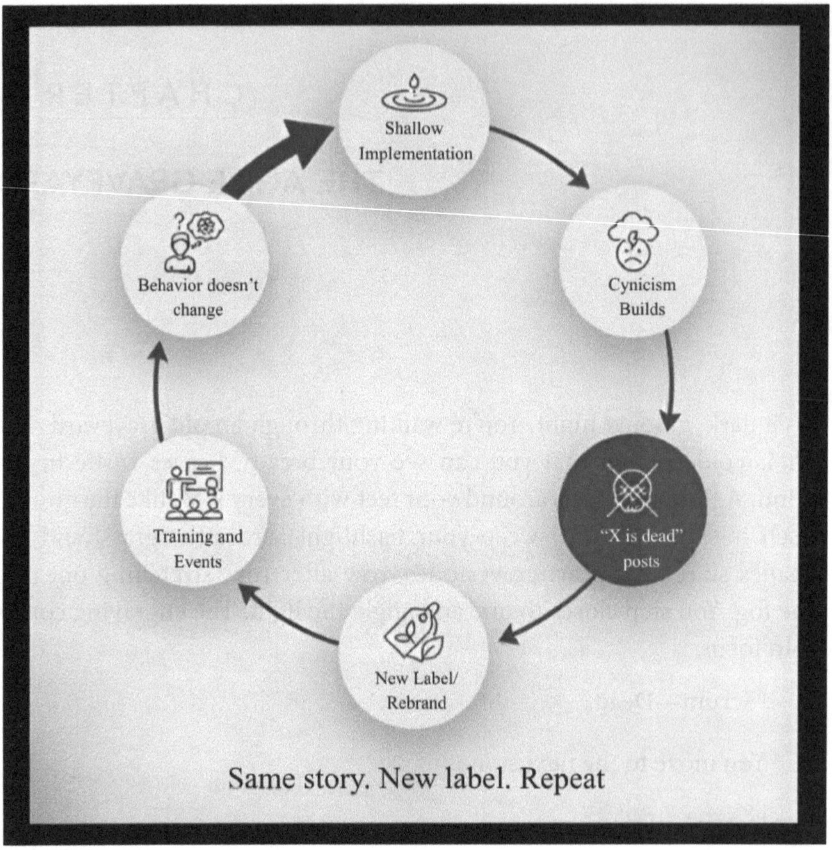

Figure 1.1 | The Agile graveyard cycle

as well build something that lasts. According to the loudest voices on the internet, this graveyard is real (see Figure 1.1). Everything is dead, useless, obsolete, or not fit for the modern world. It's a constant stream of funerals, delivered with absolute confidence and zero curiosity.

So here's the question you have to ask yourself as you stand there with that flashlight: Is any of this really dead? Or are we just addicted to declaring things dead because it gets attention, clicks, comments, shares, speaking slots, and the next wave of hype? You see the cycle everywhere: on LinkedIn, Reddit, conference stages, and in blogs. Post after post, talk after talk, someone's always declaring, "X is dead," "Y is obsolete," "Z is the only way forward now." We end up stuck in a never-ending funeral for last year's ideas, always chasing the next big thing, always ready to bury what came before. And the irony is that most of the time these things aren't dead at all. They're just out of fashion. Or someone has a new product to sell.

Meanwhile, the real work, the stuff that helps teams and organizations get better, gets lost in the noise. Here's what doesn't change just because the internet got bored: there's decades of research showing that self-managed teams can outperform traditional, tightly controlled ones; that people tend to feel and perform better when they have autonomy; that work gets done faster and with higher quality when teams are trusted and supported; and that leadership approaches beyond command-and-control aren't just trendy—they're effective. People respond to them because they're human.[1,2] Before we start digging more graves, we should pause and ask a better question: Are we learning from what works, or are we chasing the next funeral?

The need for agility isn't new. The human principles behind it—adaptation, learning, trust, and autonomy—have existed for centuries. What's new is the name, the branding, and the industry that grew around it. Agility didn't begin with 17 people at a ski resort writing the Manifesto for Agile Software Development in 2001. That wasn't the birth of agility. It was a moment where people shone a light on something that already existed. What those folks did in 2001 was courageous. They looked at the way software was being built and said, "We can do better." Their focus was software, sure, but the ideas spread because the underlying problems weren't unique to software. The same patterns show up in marketing, operations, service work, retail, healthcare, and small businesses: unclear priorities, slow decisions,

---

1 On self-managed teams and the conditions under which they outperform traditional, tightly controlled structures, see J. Richard Hackman, *Leading Teams: Setting the Stage for Great Performances* (Harvard Business School Press, 2002). On autonomy, intrinsic motivation, and performance, see Edward L. Deci and Richard M. Ryan, *Intrinsic Motivation and Self-Determination in Human Behavior* (Plenum, 1985), and later work on Self-Determination Theory, which shows that autonomy, competence, and relatedness are key drivers of well-being and effectiveness at work.

2 For research on leadership approaches that outperform traditional command-and-control and foster learning and performance, see Amy C. Edmondson, "Psychological Safety and Learning Behavior in Work Teams," *Administrative Science Quarterly* 44, no. 2 (1999): 350–383; Amy C. Edmondson, *The Fearless Organization: Creating Psychological Safety in the Workplace for Learning, Innovation, and Growth* (Wiley, 2018); and Ronald A. Heifetz and Marty Linsky, *Leadership on the Line: Staying Alive Through the Dangers of Change* (Harvard Business Review Press, 2002).

fear-based leadership, and systems that punish learning. Agility grew. It evolved. That's the whole point.

Agility isn't what's dead. What's buried here are the half-baked, cargo-cult implementations that never had a chance, because they tried to install a process without changing behavior, ethos, or leadership. But instead of learning from that, we blame the tombstone and move on to the next fad.

Agility Isn't New

I'm not going to claim agility has been around since the dawn of humankind 6 million years ago (thank you, Google), but it has been around a long time. If you zoom out and look at how we've evolved as a society, you start to see it everywhere. Any time people had to innovate, adapt, learn from feedback, or change direction when reality changed on them, they were doing the thing we now call Agile, long before anyone put it in a full-day workshop.

One of the biggest myths we've built over the last couple of decades is that an Agile environment must always produce innovation as the end goal. If it's not churning out disruptive products and viral success stories, it must be failing. But innovation isn't the endpoint. It's a byproduct. The same goes for speed. People love to talk about going faster as if that's the whole point of agility. It isn't. Speed is what you get when you remove friction and confusion from a system that's already set up to learn and adapt.[3]

The real purpose of agility is simpler and more grounded than the industry makes it out to be: adapt to your environment, respond to variables rather than pretending they don't exist, replan when new information comes in, and improve as you go rather than locking yourself into a bad plan and hoping for the best. That's the job. Not "move faster at all costs." Not "ship more." Learn, adjust, and get better, without breaking people.

You can see these principles even in places you'd assume are rigid, like the military. On paper, the military is strict and deeply hierarchical. But if you look at how effective units operated in the world wars and

3 This aligns with the Agile Manifesto's emphasis on "responding to change over following a plan" (2001) and with research in Lean and DevOps that frames speed and innovation as outcomes of healthy systems focused on flow, feedback, and learning. See, for example, Nicole Forsgren, Jez Humble, and Gene Kim, *Accelerate: The Science of Lean Software and DevOps* (IT Revolution Press, 2018).

beyond, you see something more practical than pure top-down control: small teams adapting in real time to chaos on the ground. You see commanders giving intent and boundaries, then trusting the people closest to the action to make decisions. They called it decentralized command.[4] It's not that different from what we mean when we talk about self-organizing teams. There's a clear overall goal. Teams are given intent and constraints. They're trusted to figure out how to achieve the goal. In *Extreme Ownership*, Jocko Willink frames decentralized command as a core leadership principle, and when you read the stories, something clicks: we're totally fine trusting small teams to make life-or-death decisions in chaotic, high-risk environments… but we somehow can't trust a software team to decide how to slice a feature without three layers of approval. Think about how backward that is.

On the battlefield, leaders set clear intent, teams understand the mission and constraints, and decisions get pushed down to the people closest to the action, because that's where the information is. In most organizations, leaders set vague slogans, teams get tasks instead of intent, and decisions get pulled up to the people farthest from the work, because that's where the titles are. We accept decentralized command when bullets are flying, but when it's just code, marketing campaigns, or service work, suddenly we need command-and-control. That's not a capability problem. That's an ethos problem.

And just so we're clear on what I mean by ethos: ethos is your lived values—what you actually believe and reward, not what you print on a poster. If you believe people can't be trusted, you'll build systems that treat them like they can't be trusted. You'll add layers of approval, status reports, and signoffs that quietly say, "We don't believe you'll do the right thing unless we watch you." If you believe people are capable, you'll build systems that let them prove it. You'll design for clarity, feedback, and autonomy instead of control.[5] That belief, your ethos, is what shapes the environment.

4 On decentralized command and mission-type tactics in military history, see Martin van Creveld, *Command in War* (Harvard University Press, 1985), and discussions of Auftragstaktik (mission command) in German and Allied doctrine. For a modern leadership framing, see Jocko Willink and Leif Babin, *Extreme Ownership: How U.S. Navy SEALs Lead and Win* (St. Martin's Press, 2015).

5 On how leaders' underlying assumptions about people (trust vs. control) shape the systems they build, see Douglas McGregor, *The Human Side of Enterprise* (McGraw-Hill, 1960), which contrasts "Theory X" (people must be controlled) with "Theory Y" (people are capable and self-motivated), and Edgar H. Schein, *Organizational Culture and Leadership* (Jossey-Bass, multiple editions).

Agility as an Ecosystem

This is what I mean when I talk about an Agile environment, or better yet, an ecosystem. Different language, same principles. The goal of an Agile environment isn't perfection. We're never going to get there. Perfection is a moving target, and chasing it usually makes people miserable, risk-averse, and scared to experiment. The goal is to create a work ecosystem where a few healthy practices become the norm. Ideas flow instead of getting stuck in someone's inbox for three weeks while everyone pretends they're "blocked." Work gets done in a way that feels sane, not like a constant emergency where every day is a new crisis. People don't dread Monday morning because they aren't bracing for another arbitrary fire drill. They have space to think things through instead of reacting to the latest ping, and new solutions can come from anywhere, not just from the highest-paid person in the room.

In a healthy ecosystem, new ways of working get explored, tested, and either adopted or dropped based on what works, not based on who suggested them. Red tape that doesn't serve a real purpose gets removed. Unnecessary bottlenecks and handoffs are reduced. Teams can get to the finish line without being blocked at every turn by outdated command-and-control. That was the good intention. That's what agility was meant to support. Not a specific set of events. Not a specific brand of framework. A different kind of ecosystem.

Somewhere along the way, we buried that intention under process worship, tool obsession, and a whole lot of one-true-way thinking. We made the ecosystem about the plants, the tools, and frameworks, instead of the soil: the ethos, structures, and behaviors that let anything grow in the first place (see Figure 1.2).

Later in the book, we'll dig into agility as an ecosystem; how leadership, structures, incentives, tools, and behaviors interact; and how small changes in one part can either support or suffocate the rest. For now, hold on to this: if your environment makes it hard for ideas to flow, for people to think, and for teams to adapt, it doesn't matter what you call it. It isn't Agile.

So, Why Are We in a Graveyard?

If everything is supposed to evolve, why hasn't agility? Why are we still struggling with it? Why do we keep hearing *Agile is dead*? It's not

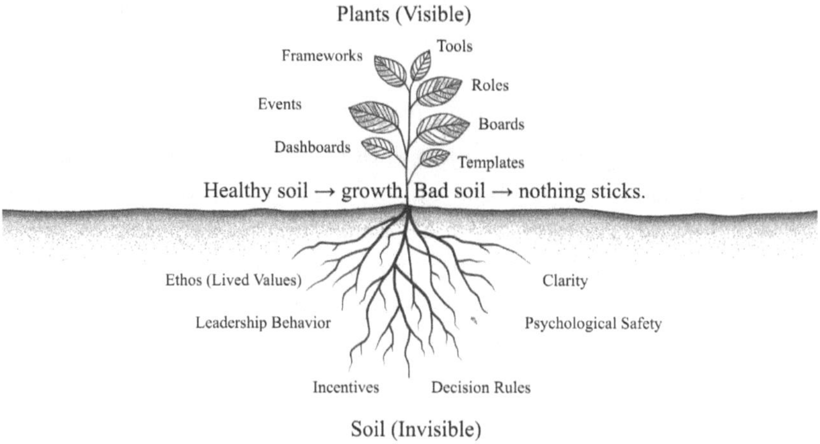

Figure 1.2 | Plants versus soil (visible versus invisible)

because the ideas are broken. It's because we've lost our way. We've let other things get in the way: leaders giving lip service to Agile and work-life balance while behaving the same old way; organizations slapping Agile labels on rigid structures; and frameworks being treated like religion rather than tools. Almost every leader can say the right buzzwords: *We're an Agile team. We care about work-life balance. We empower our people.* The harder question is: How many leaders live it when it costs them something? If you walk around most organizations and ignore the posters, you'll see a very different story—Agile teams that still need three approvals to release anything; work-life balance cultures where people get praised for answering emails at 11 p.m.; and empowered teams that get overruled the moment their decisions make someone uncomfortable. On the wall, it's Agile. In the room, it's the same old command-and-control.[6]

So we fall back into the same patterns, the same behaviors and structures that caused the problems in the first place. And in this

6 On how underlying culture and structures pull behavior back to the status quo, even when organizations claim they want change, see Edgar H. Schein, *Organizational Culture and Leadership* (Jossey-Bass, multiple editions), and Chris Argyris, *Overcoming Organizational Defenses: Facilitating Organizational Learning* (Allyn & Bacon, 1990).

graveyard, we've seen the movie before. Things "die," then come back as a phoenix... or a zombie. They get resurrected, renamed, and rebranded. Just like wash, rinse, and repeat.

We even have our own rituals: Plan–Do–Check–Act. We treat it like gospel. We treat agility like it's just another cycle to run through: plan the transformation, do the training, and check the boxes. Then we end up with Frankenstein frameworks—pieces of this, fragments of that, stitched together and sold as *the hybrid that will finally work.* You've probably seen it: a bit of Scrum here, a slice of SAFe™ there, some Kanban terms sprinkled on top, and a custom maturity model to make it look scientific. When it doesn't work, we still don't question ethos, behavior, or leadership. We don't ask what we actually believe about people and control. We don't ask what we're rewarding. We don't ask what we're afraid to let go of. Instead, we shrug and say, "Agile doesn't work here." And then we go back to the graveyard and add another stone.

The Baggage We've Created

All these failures, half-baked transformations, and endless rebrands have turned the term *Agile* into baggage. And not the good kind. Not the suitcase you bring back from a great trip, full of souvenirs, photos, and memories. This is the kind of baggage that's been loaded with rocks, heavy, unpleasant, and passed around by people who don't care what's inside. That's what we've done to Agile.

We've created an industry where the word itself is loaded. You can feel it the moment it comes up in a meeting. People roll their eyes. Teams groan when a transformation is announced. Leaders treat it like just another management fad that will blow over if they wait long enough. That's the graveyard you're walking through, not a cemetery of tools and frameworks that failed us, but a graveyard of misuse, misunderstanding, and missed opportunities.

This is a pattern you've probably seen up-close. An organization sends teams to training. Maybe it's an intense full week. Maybe it's one day a week for three months. They bring in an Agile coach. They roll out a framework. They run some sprints. They add stand-ups and retros. They put up a board on a wall or in Jira. On the surface, it looks like a lot is happening. There's motion. There are new words. There are new meetings.

People can point to activity. And then someone, usually in a leadership meeting, says the magic words: *We're an Agile organization now.*

Meanwhile, under the surface, almost nothing that matters has changed. Leadership hasn't changed how they behave. Decision-making is still top-down. Priorities are still unclear and change on a whim. People are still afraid to speak up when something's broken. The system still rewards compliance and punishes bad news. The only difference is that now the same old problems are happening inside a new set of ceremonies. But hey, teams had training. So we must be Agile... right? This is how you end up with "Agile is dead" stories. Not because the ideas are flawed, but because the implementation was superficial and misaligned with how the organization actually functions. In other words, the label changed. The behavior didn't. Implementation: dead.

The Coaching Problem (and Comfort Problem)

There are great Agile coaches out there. You've probably met a few. They're the ones who coach, not just present a deck. They sit with teams and leaders. They ask uncomfortable questions. They don't just teach people how to do stand-ups; they help executives change how they show up. They push on character and values, not just events and tools.

And then there's another pattern: coaching that stays safely at the team level. A coach runs a two-day training, hands over a slide deck, and moves on. Or they spend months tweaking boards and stand-ups because leadership is "out of scope," too political, or too hard to confront. The problem is that teams don't operate in a vacuum. You can't workshop your way out of a system that rewards interruption, avoids trade-offs, and punishes honesty. Even the best coach can't fix one core issue: many people don't actually want to change. Comfort is powerful, even when the current mess is clearly not working. What looks like resistance to change is often resistance to loss[7]. Loss of control. Loss of status. Loss of certainty. Loss of familiar routines and the stories people tell themselves about why their role matters. Change is constant; we adapt to weather, markets, and technology all

7 Ronald A. Heifetz and Marty Linsky famously argue that "people do not resist change, per se. People resist loss." See Ronald A. Heifetz and Marty Linsky, *Leadership on the Line: Staying Alive Through the Dangers of Change* (Harvard Business Review Press, 2002).

the time. What people resist is what they might have to give up in order to change.

You can hear it in the questions nobody wants to ask out loud: *If we really empower teams, what happens to my role? If we stop doing things my way, do I still matter here? If we admit this isn't working, what does that say about my past decisions?* No coach, no framework, and no tool can override that if the people in power are more committed to protecting comfort than improving the system. It's easy to blame coaching. The deeper problem is comfort. If you want a modern example, look at artificial intelligence (AI). It's the latest tool shaking up industries, and people worry it will take jobs or disrupt everything. Underneath that fear is the same old resistance to loss: control, certainty, and familiar ways of working. And this isn't new. In the early to mid-1990s and into the 2000s, software testing underwent a similar shift toward test automation.[8] I was working with testing groups, leading testing projects and programs, and the fear was intense: *If we don't learn automation, we'll lose our jobs. Automation will replace testers.*

My argument then is the same one I'd make about AI now. Yes, automation replaces repetitive, low-value tasks. Yes, it handles regression and repeatable checks faster than a human ever could. But that's the point. Used well, automation frees people to be more valuable, not less. It creates room for more strategic thinking about risk, better test design, deeper collaboration, and better judgment about what matters and why.

AI is no different. Used well, it can help us do things better, reduce grunt work, and force us to level up our thinking and skills. Used poorly, it amplifies existing dysfunction, accelerates poor decisions, and makes people feel even more disconnected and threatened. The pattern repeats: a new tool appears, fear spikes, and people use it to either evolve or reinforce the same old problems. It's not the tool. It's the implementation. The same applies to agility. Here's an analogy that still fits. Imagine a zoo putting a jaguar in a polar bear enclosure. Technically, the jaguar will survive. It can breathe, move, and eat. But it's going to be miserable.

8 On recurring fears about automation and job loss, and how technology tends to reshape work rather than simply eliminate it, see James Bessen, *Learning by Doing: The Real Connection between Innovation, Wages, and Wealth* (Yale University Press, 2015), and Erik Brynjolfsson and Andrew McAfee, *The Second Machine Age: Work, Progress, and Prosperity in a Time of Brilliant Technologies* (W. W. Norton, 2014).

Wrong climate. Wrong terrain. Wrong everything. The problem isn't the jaguar. The problem is the environment.

That's what a lot of Agile transformations feel like. You take a team and slap Agile language and events (stand-ups, retros, boards, and sprints) on them while keeping everything else the same. The rigid hierarchy stays. Command-and-control behavior stays. Fear stays. On the surface, it looks like something changed. Underneath, nothing did. You've put agility, the jaguar, into a hostile environment and then blame the jaguar when it doesn't deliver. And bad implementations aren't neutral. They're cruel. They convince people that agility itself is the problem, when the enclosure was never designed to support it in the first place.

Whether you're building an Agile environment, introducing a new process, adopting a new tool, or rolling out AI and automation, it comes down to one thing: behavior change. If behaviors don't change, nothing else really does. The ethos stays the same. The values stay the same. The way decisions get made stays the same. And if those don't change, your implementation is dead on arrival, no matter how many new terms or tools you throw at it.

To make agility work, really work, you have to change how leaders behave, how teams interact, how decisions are made, and how you respond to uncertainty and feedback. That's the difference between a story that ends with "Agile is dead" and a story that ends with "Our implementation was shallow, misaligned, and never had a chance." Let me give you a real example of how this plays out. I once worked with an organization that said they wanted to "be more Agile." Not just in name, they wanted to be more adaptive, more efficient, and more effective across the whole organization. So we started the right way: with an assessment. We looked at how work flowed, how decisions were made, how teams interacted, and where the real bottlenecks and pain-points lived. From that, I built recommendations to support real agility across the system: structural changes, changes in leadership behavior, and new ways of working that would make it easier to adapt and respond, not just for a single team but across the whole organization. On paper, they agreed. In practice, what they really wanted was something else: Just train the teams.

What they meant was: teach teams how to "be Agile." Show them techniques and tools. Help them run events, use boards, estimate, slice work, and build better habits. But don't touch the rest of the system. Don't change policies. Don't change decision-making. Don't change leadership behavior. Don't change what gets rewarded. The plan was straightforward. There were eight teams. We'd work with two teams at a time over roughly three-and-a-half months per pair, meeting weekly with each team to teach techniques, tools, and habits meant to improve adaptability and flow. My intent was twofold: help the teams genuinely improve their work and, in turn, show leadership what else needed to change beyond training, including structures, policies, decision patterns, and their own behaviors.

Here's what happened. After the first two teams completed their cycle, the organization stopped the transformation. No work with the remaining six teams. No bigger organizational changes. No follow-through on assessment recommendations. They got some training, then pulled the plug. And what happened next is exactly what you'd expect: the teams gradually reverted. The new techniques were forgotten or used only partially. The environment around them hadn't changed, so the change couldn't stick. This wasn't a failure of the teams. It wasn't a failure on the part of the people trying to help. It was a failure of implementation, driven by leadership's fear of loss. Leadership wanted the benefits of agility without changing the system that made agility impossible. They feared losing control, status, and a familiar way of leading. So the transformation died. Not because agility doesn't work. Not because the teams couldn't handle it. It died because leadership chose comfort over change.

When you're walking through the Agile graveyard, it's tempting to blame the tools. It's easy to say, *Scrum failed us*, *SAFe™ doesn't work*, *coaches didn't deliver*, or *the teams weren't ready*. That story is comforting because it lets us point to something outside ourselves. But if you look closely at most of these stories, the pattern usually isn't that Agile is broken. It's closer to this: we tried to change everything except how we behave. We changed terminology, but not how decisions get made. We added events, but didn't make it safe to tell the truth. We introduced boards and tools, but didn't create clarity or trust. We trained teams but never touched leadership ethos. That's not an agility problem. That's a behavior problem.

When implementations fail, it's almost never because a framework is inherently flawed or a team is incapable of learning new techniques. It's because leaders won't relinquish control. It's because people are more afraid of loss than of committing to growth. It's because the organization's character and values stay the same. You can't build an adaptive environment on fear, ego, and control with a thin layer of empowerment slogans on top.

So if agility isn't dead, and I don't believe it is, and if the tools aren't the real issue, and if the graveyard is mostly full of bad implementations, then we have to stop asking: *Which framework will save us*? and start asking harder questions: *What behaviors are we rewarding, what do we value when things get messy, and who are we, leaders and teams, when things get hard?* Frameworks aren't the enemy. The way we cling to them, sell them, and weaponize them is.

That's where we're going next. In Chapter 2, we'll dig into The Framework Fiasco: how frameworks turned into religion, why "by the book" so often goes off the rails, and what changes when you stop treating a framework as your identity and start treating it as one tool in a much bigger toolbox. If you're willing to examine your relationship with frameworks and your need for certainty and labels, turn the page.

CHAPTER 2

THE FRAMEWORK FIASCO

At the end of Chapter 1, we landed on something you need to keep top of mind: most "Agile is dead" stories aren't about broken ideas. They're about behavior, ethos, and what leaders reward, tolerate, and protect when things get uncomfortable. But before we can get anywhere useful, there's another layer to this mess that we need to talk about: our relationship with frameworks.

Why are we so loyal to one framework? Why do people treat Scrum versus SAFe™ versus "our custom hybrid" like a holy war instead of what it really is—a choice of tools? Over the past two decades, as Agile frameworks exploded in popularity, something weird happened. They stopped being tools and started becoming identities. You've heard the phrases: "We are a Scrum shop." "We do SAFe™ here." "This framework is the right way." Not "this is useful for us right now," but "this is the way." And yes, that's a Mandalorian reference. My inner geek refuses to sit this one out.

Even Scrum had to pull itself back from the edge. When the Scrum Guide was updated around 2020, the language shifted from *ceremonies* to *events*.[1] That might sound like a minor edit, but it matters. Ceremonies sound like ritual. Ritual turns into habit without thought. Habit without thought turns into dogma. Events are just that: things you do for a reason. When people start defending the ritual long after they've forgotten its purpose, you no longer have a framework; you have a belief system.

1 The 2020 update to the Scrum Guide removed the term "ceremonies" and consistently refers to "events," part of an effort to make Scrum less prescriptive and more focused on purpose and outcomes. See The 2020 Scrum Guide, available at scrumguides.org.

Frameworks were supposed to be tools. They were never meant to be the end goal. The end goal is much simpler and much more human: achieving outcomes, delivering value to clients and customers, improving how your organization works, and making work more effective and humane. The question is never "What is the right framework?" The real question is: *Is this the right tool for this team, in this context, to deliver the value we need?* That's a very different conversation.

If you ask most people to name an Agile framework, you'll usually hear Scrum or SAFe™. They're the biggest and loudest. But if you do a quick search for "Agile frameworks," you'll find a whole zoo of them. A lot of them exist because someone got frustrated with the mainstream options and built an alternative. Fair. The problem is what happens next: over time, that alternative becomes its own little identity. "This is the best framework." "This is the one that finally works." No. It might be the best framework for that team, in that context, at that time. That doesn't make it the best for everyone.

You can argue that Scrum is used by most of the market, often quoted in the 70–80% range, depending on which survey you look at.[2] Fine. That tells you what's popular. It doesn't tell you what's right for your situation, or whether anyone using it understands why they're doing what they're doing. That's the fiasco we're about to unpack. Because the real question isn't how many people are using Scrum or SAFe™. The real question is whether they're getting value from it, or whether they're just swinging a hammer at everything, including screws. If you give someone a hammer and teach them how to use it, that's great, as long as the job involves nails. If the job requires screws, that hammer will still get something into the wood, sure, but it won't be pretty, stable, or built to last (see Figure 2.1).

Call it religion, call it a cult, or call it strong preference, I honestly don't care what label you use. The point is simple: frameworks are tools. If you forget that, you're already in trouble.

2 Various industry surveys consistently report Scrum as the most widely used Agile framework, often in the 70–80% range among respondents. For example, see the 15th State of Agile Report (Digital.ai, 2021) and similar annual surveys.

Figure 2.1 | Use the right tool for the job. Many organizations picked a hammer, and now they're trying to solve every problem with it

The Damage of Process Obsession over Outcomes

This obsession with frameworks and processes is a big reason why so many Agile transformations flop. You don't need another statistic to prove it; just look around. The pattern is everywhere: transformations promise better outcomes and deliver a lot of activity instead. The exact number doesn't matter nearly as much as the repetition of the cycle.[3]

So why does it keep happening? Because organizations fixate on process while ignoring the factors that determine whether change actually sticks: behavior, culture, values, and outcomes. Leadership says, "We'll implement Scrum/SAFe™/whatever, and that will make us Agile. We'll roll out this framework, and things will get faster." What they don't say is, "We'll change how we lead. We'll change how we make decisions. We'll change how we treat people." So attention goes to roles, events, artifacts, and tooling. People obsess over whether the stand-up is exactly 15 minutes, whether the board is in Jira or on a wall, and whether the backlog is refined properly, while the questions that actually matter get

3 Various studies on organizational and digital transformations report high failure rates, often around 70%. See, for example, McKinsey & Company, "Perspectives on Transformation," which notes that "Seventy percent of transformations fail," and Boston Consulting Group, "Flipping the Odds of Digital Transformation Success" (2020), which reports that 70% of digital transformations fall short of their objectives.

sidelined: *Is any of this producing better outcomes? Are people burning out? Is the product better? Are customers getting more value?*

Sure, you might get a product out the door. You might even get a bit more speed than before. But is it the right product? Is it the most valuable thing you could be delivering? And is it sustainable for the humans doing the work? If the answer to those questions is "no," then it doesn't matter how perfectly you're doing the framework. You optimized the process and missed the point. Process obsession blinds you. When teams understand why they use specific tools and events, what outcomes they're aiming for, and when the organization is aligned on value and flow, the process naturally evolves. It bends to reality rather than forcing reality to bend to it. When you're fixated on doing the process right, the opposite happens. You can end up creating value at the expense of the people making it. You burn people out and keep obstacles in place. And you lose sight of whether any of this is actually helping.

When you focus on outcomes rather than rituals, you become more willing to adapt the process. You drop what doesn't work. You experiment more. And you're far more likely to protect the humans doing the work, not just the metrics on the slide. That's agility, not "we followed the framework perfectly."

Frameworks are tools in a toolbox. Scrum is a hammer. Kanban is a screwdriver. SAFe™ is a massive, complicated toolset. Extreme Programming (XP) is closer to precision instruments. None of them is "the way." They're just different tools for different jobs. And sometimes the best results come from deliberately combining complementary tools. XP practices, when used alongside Scrum, can improve technical quality and team collaboration. Within SAFe™, there are different ways to create and manage requirements depending on the context. Mixing and matching isn't the problem. The problem is mixing and matching without understanding what each thing is for, bolting together pieces of every framework you've ever heard of, and hoping a Franken framework will magically fix everything (see Figure 2.2).

That's how you end up with a monster: all the rules, all the jargon, all the overhead, and none of the outcomes. You get motion without progress. You get busy without actually getting better. A good

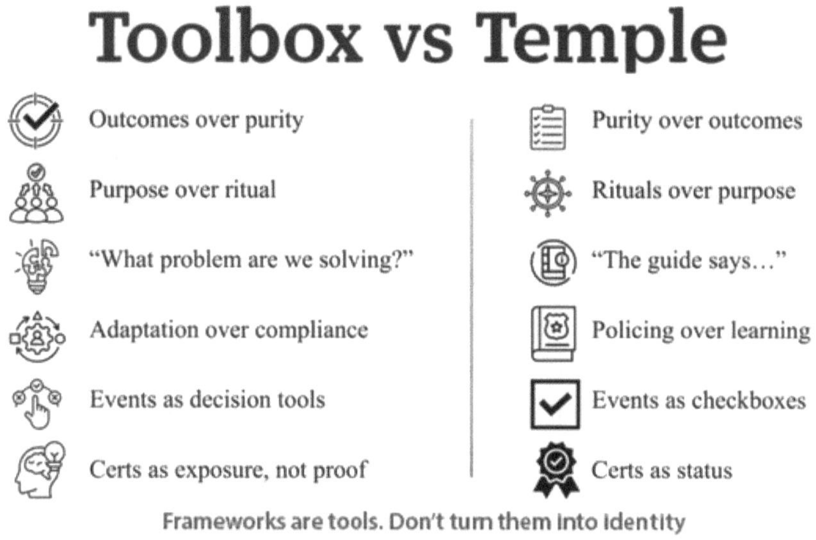

Figure 2.2 | Tools, not temples

craftsperson doesn't grab every tool and start swinging. They know what each tool is good for, what it's terrible at, and when to put it back in the box. Sometimes you need a hammer and a screwdriver. Sometimes you need to invent something new. But you always start with the problem, not the tool.

Frameworks should work the same way. The real question isn't, "Are we doing Scrum/SAFe™/Kanban properly?" It's, "Given this team, this context, and this outcome, what combination of tools, techniques, and practices actually makes sense, and what should we leave on the shelf?" When you treat a framework like a temple, you protect the rituals. When you treat it like a toolbox, you protect the people and the outcomes. A lot of organizations think they're using a toolbox, but they're really just following rituals without understanding why. That's how you end up with the worst of both worlds: more process, less learning, and a team that feels like they're failing at something they were never set up to succeed at. Real agility is making intentional choices. It's adapting techniques to fit the work and the humans doing it, not following a script or worshiping the tool.

When the Framework Becomes a Performance

So how do you know when you're stuck in a performance instead of building real agility? The signs are usually obvious once you know what to look for. The common thread is simple: you're doing the motions, but you're not getting the value.

Events Become Rubber Stamps

You run the events exactly as the guide recommends: daily Scrum (or daily stand-up), sprint planning, review, and retrospective. On paper, you're "doing it right." But over time, those events can turn into checkboxes—"Yep, we did it"—and slowly become time sinks with no real value. People show up because they're supposed to, not because the event helps them make decisions, solve problems, or deliver better outcomes.

The questions you should be asking are simple: *What is the expected value of this event, and did we actually get that value?* If you can run a daily in five minutes and people leave with clarity, you delivered value. But if you sit in a 15-minute daily meeting and no decisions are made, no help is requested, and no clarity is created, you're not building agility. You're acting it out.[4]

Retrospectives That Go Nowhere

Retrospectives are one of the easiest places to spot this problem. A common anti-pattern is the classic "do well/do better." Years ago, a mentor of mine remarked, "do well/do better" is one of the most useless patterns you can use in a post-mortem. That was in a Waterfall context, and it's just as ineffective here. Think about it. "What did we do well?" "What should we do better?" Then what? Did you track what you did well and make sure you kept doing it? Did you turn "do better" into actual experiments, actions, and follow-ups? Most of the time, the answer is no.

4 The idea that Agile events should serve a clear purpose rather than exist as empty rituals is consistent with the Agile Manifesto's focus on "individuals and interactions over processes and tools" and with guidance from experienced practitioners. See, for example, Esther Derby and Diana Larsen, *Agile Retrospectives: Making Good Teams Great* (Pragmatic Bookshelf, 2006), who emphasize designing meetings and retrospectives around explicit goals and outcomes rather than rote formats.

You end up with vague statements like: "We should communicate better," "We should write better test scripts," or "We should estimate better," and then... nothing. In an age where information is everywhere and we can track, measure, and learn it in ways we couldn't before 2000, it's wild how many retros still collapse into fuzzy statements with no follow-through.[5] That isn't agility. That's a meeting that makes people feel busy.

The Process Matters More Than the People

Another sign you're stuck is when people are clearly frustrated or burned out, obstacles persist, and the main concern in leadership meetings is still, "Are we doing the framework correctly?" If your primary question is, "Are we following Scrum/SAFe™/Kanban properly?" instead of "Are we sustainably delivering value?" then you're protecting just the script, not the humans doing the work.[6]

If you're wondering whether your framework has turned into a performance, here's a quick self-check. Be brutally honest:

1. **Events:** If you canceled your daily stand-up, planning, or retro for a week, then what would actually break? Would outcomes and clarity suffer, or would it just feel wrong because "that's not how we're supposed to do it"?
2. **Decisions:** During your events, are real decisions being made and acted on, or are you mostly doing status updates and recycling the same complaints?
3. **Retros:** Can you point to specific changes you've made in the past three months that came directly from retrospectives? Are you tracking whether those changes helped?

---

5 On using retrospectives as genuine learning and improvement mechanisms rather than vague complaint sessions, see Esther Derby and Diana Larsen, *Agile Retrospectives: Making Good Teams Great* (Pragmatic Bookshelf, 2006), and Norm Kerth, *Project Retrospectives: A Handbook for Team Reviews* (Dorset House, 2001). Both stress the importance of concrete actions, experiments, and follow-up as opposed to generic "we should do better" statements.

6 The emphasis on people and sustainable value over strict process adherence echoes core Agile principles such as "individuals and interactions over processes and tools" and "working software over comprehensive documentation" from the Agile Manifesto (2001), as well as the principle that "Agile processes promote sustainable development." See also Jeff Sutherland and Ken Schwaber's *The Scrum Guide* (various editions), which frames Scrum as a framework for generating value, not an end in itself.

4. **People versus process:** When something goes wrong, is your first question "Did we follow the process?" or "What got in the way of delivering real value, safely and sustainably?"
5. **Purpose:** If you asked your team, "Why do we run this event, use this artifact, or follow this rule?" would they give you a clear answer, or would they say, "Because Scrum/SAFe™ says so?"

If most of your answers make you uncomfortable, then you're not alone. It's rarely about picking the wrong framework. More often, it means the framework has become a script, one that overshadows the real purpose. It means you've started performing it. Frameworks aren't the enemy. Treating them like religion is. When you obsess over process, ignore outcomes, run events as rituals, and quietly accept performance instead of real value, you don't have an Agile problem. You have a framework fiasco. The way out is the same way out we talked about in the previous chapter, and what we'll keep digging into next: prioritizing behavior, ethos, outcomes, and people over process, for real, not just as words on a poster.

Scrum by the Book... and Still Failing

I've seen teams do Scrum by the book and still fail miserably. One team had everything set up exactly as the framework specifies: daily stand-ups, sprint planning, reviews, retrospectives, all the "right" roles, a clean Jira board, velocity charts, burndown charts, and the whole package. If you looked at them on paper, then you'd think: a textbook Agile team. It was a mess. Priorities changed constantly. The team would be halfway through a sprint when something urgent would suddenly appear. A leader or stakeholder would declare it a priority, and the team would pivot mid-sprint. The punchline? Most of those urgent items weren't urgent at all. They could have waited. Some should never have been in that sprint in the first place.

Micromanagement was everywhere. The team was consistently second-guessed. The Product Owner role turned into a control tower, not a value role. They believed they had to understand and control everything, which meant the team had very little real autonomy. Overcommitment became normal, too. Sprint after sprint, the team took on way too much work. It was never sustainable, but it became expected. "We'll just push a bit harder this time" turned into a way of life. It was the classic gambler's fallacy: if we just put in a little more effort, then we'll finally catch up, without noticing the odds stacked against us from the start. And, of course, burnout showed up. People were exhausted. The environment didn't feel safe. There wasn't clarity. There wasn't trust. There was pressure, more meetings, and a constant sense of being behind.

On paper, they were Agile. In reality, they were using a Waterfall approach with short iterations and additional ceremonies layered on top.[7] That's framework performance. And that's the point: you can tick every box in the Scrum Guide, run every event, use all the right artifacts, and hand out all the proper titles, and still not be Agile. Because agility isn't the framework. It's how you handle changing priorities. It's how you treat people. It's how safe it is to speak up. It's how honest you are about capacity and outcomes. Just because you're doing a framework doesn't mean you're being Agile.

There's another layer to the framework fiasco that we don't talk about enough: framework identity. This is where your job quietly becomes your religion. "I'm a Scrum Master" turns into "I must defend Scrum at all costs." A SAFe™ consultant can't admit that SAFe™ might not fit every context. A coach feels personally attacked any time someone questions their favorite model. When your identity gets tied to a framework, something dangerous happens: you stop serving the team and start serving the framework.[8]

You can see it everywhere. Coaches become more loyal to the guide than to the people in front of them. Consultants become more loyal to their certification than to the actual problem. Leaders become more loyal to "how we do Scrum/SAFe™ here" than to outcomes and

---

7 The pattern of running sprints and Agile ceremonies while still managing work in a sequential, phase-driven way is a known anti-pattern sometimes described as "Waterfall in sprints" or "iterative Waterfall." See, for example, Mike Cohn's "An Iterative Waterfall Isn't Agile," Mountain Goat Software, and discussions of "Waterfall in sprints" in Agile anti-pattern guides such as ProdPad's "11 Agile Anti-Patterns Product Managers Need to Watch."

8 On the risks of over-identifying with roles, labels, and models, see Chris Argyris's *Overcoming Organizational Defenses: Facilitating Organizational Learning* (Allyn & Bacon, 1990), which explores how professionals protect their own competence and status instead of confronting problems; and Edgar H. Schein's *Organizational Culture and Leadership* (JosseyBass, multiple editions) discusses how deeply held assumptions and identities can block learning and change.

well-being. If your first instinct is to protect the framework rather than help the team, that's a signal. It's worth pausing and asking what you're truly loyal to. We need to move away from the framework-loyal coach and toward the outcome-loyal coach. The one who starts with better questions: *What problem are we trying to solve, what outcome are we aiming for, and what's the lightest tool or structure that helps us get there?* That's an entirely different way of working.

Let's go back to the toolbox analogy, but take it a step further. Think of a basic toolbox. Scrum is a hammer. Kanban is a screwdriver. SAFe™ is a full power toolset. XP is a precision tool. But your toolbox isn't just frameworks; it's also techniques and practices. Stand-ups are a measuring tape. Retrospectives are a level. Work-in-progress limits are a clamp. Pair programming is a specialty wrench. None of it is the way. It's all just different tools for different jobs.

I had an uncle who was a carpenter. He wasn't married to one tool. He didn't say, "I'm a hammer guy, so I use a hammer for everything." He used hammers, screwdrivers, saws, levels, power tools, whatever the job required. Sometimes he'd even make his own tool if nothing else fit. He used everything at his disposal to deliver great work and real value to his customers and himself. He even built his own house that way. He didn't worship the tools. He used them. That's how we need to think about frameworks and techniques. Pick what fits the job. Adjust when reality changes. And sometimes, yes, you may need to switch tools mid-job—not in the middle of a sprint, because that's chaos, but when the work shifts or when you realize what you're using isn't cutting it anymore. At that point, the question isn't "How do we follow the guide more strictly?" It's "Is this tool still helping us, or is it time to try something else?"

That shift changes everything. You move from "We must do it this way" to "What's the best way to support this team and this outcome right now?" Until we make that shift, we'll keep seeing the same pattern: frameworks treated like temples, coaches treated like priests, and teams treated like followers. And we'll keep wondering why things feel rigid, fragile, and exhausting, no matter how Agile it looks on paper.

If you answer "yes" to any of the following, then you're probably stuck in framework performance:

- You argue more about what the guide says than what the customer needs.
- People are afraid of adapting the process, even when it's clearly not working.
- Retrospectives rarely lead to real change.
- Leadership asks, "Are we doing Scrum/SAFe™ right?" more than "Is it working better?"
- Teams feel more policed than supported.
- Success gets measured in events completed instead of outcomes improved.

If any of that sounds familiar, you're not broken. You're dealing with a framework fiasco. And the fix isn't more framework. It's stepping back and asking better questions: *What's getting in the way, what's the root cause, and where are we clinging to ritual instead of value?* Those questions don't get answered in a guide. They're answered by honestly looking at behavior, ethos, and the ecosystem you've created.

Certification Culture: Fuel for the Fire

A big part of why framework religion gets so strong is certification culture. There are certifications for everything now. They're not inherently bad, but they can easily reinforce wrong things. You pay for a badge, and suddenly you feel you have to defend that badge. You get certified in a framework, and everything starts to look like a nail for that hammer. The more your career and status are tied to that logo, the harder it gets for you to admit, "This might not be the right tool here."[9]

9 On the risks of credentialism and overreliance on certifications as proxies for competence, see discussions such as "Real-World Problems with Agile Certifications" by R. G. Arun (LinkedIn, 2024) and "What Is that Agile Certification Really Worth?" in *CIO* (2017) by Sharon Florentine, which highlight how certifications can create a culture where badges are valued over experience and judgment.

And it creates predictable patterns. Gatekeeping: "You're not certified, so you don't get it." Consider job postings that say "must have PMP/CSM/XYZ" while ignoring people who've done the work for years without the certification. Leaders assume "certified = competent" and "uncertified = not ready." The real risk is simple: overconfidence in the framework, and underconfidence in context and judgment. That's how certification culture feeds framework religion, which then spreads throughout the whole organization. To be clear, I'm not anti-certification. I have certifications. I've earned them. I'm anti-certification as identity, and anti-certification as proof of wisdom. A certification can show you've studied something. It can show you've been exposed to certain ideas and vocabulary. It does not prove you can lead, coach, think, or adapt in the real world.

The Real Question

Instead of asking "Which framework should we follow?" ask better questions: *What problem are we trying to solve, and what's the lightest structure that can help us solve it together?* That's an entirely different conversation. And to answer those questions honestly, you need something no framework can give you out of the box: judgment, context, and the willingness to change your own behavior.

Frameworks can help. They really can. But without judgment and context, all you'll get is more meetings, more performance, and more frustration. You don't fix a framework fiasco by buying more frameworks. You fix it by being honest about what's in the way and refusing to hide behind process when outcomes and people are clearly suffering.

And if we stopped there, this would already be hard enough. But there's another force pouring gasoline on the fire: an entire ecosystem built around selling you the answer. Influencers, gurus, and snake-oil sellers have turned agility into a marketplace of branded solutions and quick fixes. They promise transformation in 90 days, sell you the "one true way," and quietly reinforce framework religion and certification culture. When their promises don't pan out, they rarely question the method. They blame you.

As we turn the page to Chapter 3, we're going to shine a light on that ecosystem: who these players are, how their tactics muddy the waters, and how to spot the difference between genuine help and well-packaged nonsense. Because if you don't understand the game they're playing, then it's easy to mistake marketing for wisdom and end up with yet another failed Agile transformation to bury in the graveyard.

CHAPTER 3

INFLUENCERS, GURUS, AND SNAKE OIL

The internet is an incredible tool. It's also a factory for confusion if you're not paying attention. Since around 2010, there's been a massive rise in Agile influencers: coaches with big followings on LinkedIn, YouTube channels preaching "the right way," and personal brands built around "fixing Agile." Your feed is full of carousels, hot takes, and "top 10 things you're doing wrong" threads. Everyone has a take. Everyone has a template. Everyone has a "simple framework" that will change your life by Friday.

Some of these people are genuinely good. They know how to help organizations change. They respect context. They care about people, not just frameworks. They do solid work, and they make the space better. But a lot of them are selling you something you don't need, or something you already know, repackaged as their "unique method." They rebrand old ideas, slap on a new label, and sell them back to you as if they invented them. And that's where the damage starts.

How Bad Influencers Poison the Well

There's been so much shallow guidance and low-effort transformation work sold under the Agile banner that a lot of people now have a very understandable reaction:

> "We tried Agile. It didn't work."
>
> "Agile coaches are useless."
>
> "Agile transformations are a scam."

I don't blame them. If you've been burned once by a bad coach or a shallow transformation and then tried again with a different coach and got burned again, then you're not lining up for a third round. So people stop trusting coaches. Leaders stop trusting Agile language. Teams roll their eyes the moment they hear, "We're going Agile."

Meanwhile, some influencers see that pain and confusion as an opportunity. They start preying on struggling practitioners. They sell "reinvent yourself as a new kind of coach" programs. They offer "exclusive models" that are really just rebranded basics. They charge thousands for what's essentially reframing and confidence-building that could be done in a few honest conversations. Helping people grow is great. Staying current matters too. If you're not learning, you will fall behind, and you'll end up competing on price. But there's a line between professional development and identity churn, between learning new tools and being sold a brand-new personality every six months because fear converts better than nuance. That's the ethics problem hiding under the marketing.

A good coach already has what they need in their toolbox. They know how to work with leaders, VPs, executives, and business owners. They know how to adapt rather than cling to a framework. They know how to listen, observe, and help people change behavior. They don't need a new identity every six months. They need judgment, reps, and the courage to tell the truth, especially when it's uncomfortable. A lot of what's being sold these days is just old ideas with a fresh coat of paint. Same principles, same tools, same concepts, just dressed up in new acronyms, flashy diagrams, and "proprietary" language. Then it gets pitched as:

"The secret ingredient you've been missing."

"The edge other coaches don't have."

"The model that finally proves your worth."

The problem is that it adds noise. It confuses leaders, overwhelms teams, and exploits practitioners who are worried about staying relevant.[1] It also creates a culture where people think they need to buy

1 The pattern of repeatedly rebranding existing management ideas into new "models" and "frameworks" has been widely discussed in the literature on management fashions and fads. See Eric Abrahamson, "Management Fashion," *Academy of Management Review* 21, no. 1 (1996): 254–285, and related work on how "new" management concepts are marketed and diffused as must-have solutions, often without substantial novelty.

into a specific model to prove they're valuable, organizations think they need to license something special to be "real Agile," and everyone ends up chasing the next shiny thing instead of getting better at the basics. You don't need to pay a subscription fee for a premium model to demonstrate your value. If you've got a team of five delivering 10 units of value, widgets, features, whatever a week, and you help them sustainably deliver 12, you can show the uplift, the ROI, and the impact with basic math and honesty. No trademarked model required. There are plenty of free tools and straightforward ways to demonstrate value. You don't need a $X/month platform or a $2,500 value proof course to do it.

When Organizations Fall for the Show

It's not just individuals who get sucked in. Organizations do too. Leaders see slick decks, big promises, and case studies that sound amazing but are light on real detail. So they buy a branded "Agile operating model." A "90-day transformation program." A "proven framework" that guarantees results. It looks professional. It sounds confident. It feels safer than doing the slower, messier work of changing behavior. Then what happens? The consultant or influencer leaves. Nothing much changes on the ground. Teams are confused, tired, and more cynical than before. And the conclusion becomes: "Agile doesn't work here." No. That didn't work there.

I've seen teams get abandoned by this kind of shallow advice. One team I worked with had been told, "You're an Agile team now. You've learned Scrum. You're doing sprints. You're doing all the events. You're Agile." On paper, they were set. In reality, they were set up to struggle. They didn't have the support they needed. The environment around them hadn't changed. Communication patterns were still broken. Behaviors and expectations hadn't shifted. All the same old issues were still there, now with extra meetings layered on top. When they began to struggle, the coach at the time gave them one piece of advice: "You need to collaborate better." And then walked out the door. No exploration of behaviors. No attempt to understand communication styles. No work on psychological safety. No help with conflict or clarity. "Collaborate better," and gone. That's not coaching. That's abdication.

And it's not just ineffective; it's unethical. If you take someone's money or their trust, you don't get to leave them with a slogan and a shrug. Collaboration isn't a switch you flip. It's the result of trust, clarity, shared goals, psychological safety, and practical communication and conflict resolution skills. You can't just tell a team "do better" and call it a day. But this is what some teams are getting from people who call themselves coaches or influencers. And every time it happens, it harms the team, undermines trust in agility, and makes it harder for effective coaches to be taken seriously.

All of this—the rebranding, the snake oil, the shallow advice, the overpriced courses—is another reason I'm writing this book. I'm tired of seeing influencers build huge audiences by selling the same ideas with new labels. I'm tired of practitioners being told they're not enough unless they buy into the next model. I'm tired of leaders getting dazzled by buzzwords instead of doing grounded, honest work that actually changes outcomes and protects people. And the timing matters. Agility is needed now more than ever. Change has always happened, but today the pace, visibility, and interconnectedness of change are accelerating. Markets shift. Technology leaps forward. Whole industries get reshaped while you're still arguing about story points. If we don't build real agility and instead keep following false prophets and shiny models, we won't get where we need to go. We'll keep burning people out, wasting money, and adding more headstones to the Agile graveyard.

Stop shopping for answers. The real questions are: *What problem are we solving, what's the simplest structure that helps us solve it together, and how do we need to behave when it gets hard*? This includes when priorities collide, when someone powerful wants an exception, when the numbers are ugly, and when the truth is inconvenient. You can't outsource that to an influencer, a model, or a playbook. Those can give you language and tools, but they can't give you judgment, courage, or integrity. That part is on you and your leadership team. That's where we're headed next.

Let's make this practical. If you're talking to an "expert," "coach," or "thought leader," and you notice a few of these signs, your snake oil radar should start buzzing (see Figure 3.1).

SNAKE OIL SPOTTER CHECKLIST RED FLAGS

- [ ] Promises guaranteed results fast ("90-day transformation")
- [ ] Acts like the model is secret or proprietary
- [ ] Talks more about their brand than you
- [ ] Pushes high-ticket before questions
- [ ] Uses vague, magical case studies

0–1: Probably fine 2–3: Be cautious 4+: Walk Away

Good help asks questions. Snake oil sells certainty

Figure 3.1 | Snake oil spotter checklist

You might be dealing with a guru or snake-oil seller if they:

- **Promise guaranteed results in a fixed time.**

  "90-day transformation."

  "6-week culture reset."

  "Double productivity in 30 days."

- **Act like their model is proprietary and secretive.**

  "I can't explain it fully here—you have to take the course."

  "It's too advanced to share in a post."

- **Talk more about their brand than your context.**

  "My method." "My framework." "My system."

  Very little about your reality.

- **Push high-ticket programs before understanding your situation.**

  They don't ask many questions.

  They jump straight to "You need my premium program."

- **Dismiss your concerns with labels instead of curiosity.**

  "You're just resisting change."

  "That's your fixed mindset talking."

  Anything that shuts down the conversation instead of engaging with it.

- **Use vague, magical case studies.**

  "We improved productivity by 300%."

  No baseline. No context. No trade-offs. No details.

If two or three of these show up together, you're not in a conversation about help. You're navigating a high-pressure sales pitch. By contrast, the people worth listening to usually do the opposite. They ask a ton of questions about your context. They're clear about what's new versus what's borrowed. They're honest about where their approach doesn't fit. And they don't need to hide everything behind a paywall to sound valuable.

Thought Leadership Versus Thought Marketing

Not everyone with a big following is a villain. But not everyone with a big following is a leader, either. There's a real difference between thought leadership and thought marketing (see Figure 3.2).

Thought leadership shares ideas openly, not as bait for a high-pressure sales pitch. It credits the people and research from which those ideas came. It acknowledges nuance, trade-offs, and limits. It says, "It depends," and then actually explains what it depends on. It gives you enough detail to try something without paying first. Thought marketing does the opposite. It teases "secrets" you can only get in the course. It oversimplifies complex problems into catchy slogans. It frames everything as "you're missing this one thing." It makes you feel behind, so you'll buy the fix. It hides behind IP instead of being transparent. One is trying to advance the field. The other is trying to move you into a pipeline.

Figure 3.2 | Thought leadership versus thought marketing

I want to be clear: I'm not saying people can't sell things. I sell things, and you probably do too. There's nothing wrong with making a living. The problem is when marketing replaces substance, hype replaces honesty, and "brand" replaces experience and results. Here's a simple check: if you come away from someone's content feeling informed, challenged, and clearer about what you could try next, you're probably looking at thought leadership. If you come away feeling anxious, inadequate, and like you're missing some ingredient only they have, you're probably looking at thought marketing.

Inside the Influencer Playbook

Let me walk you through a pattern I've seen more than once. A mid-level Scrum Master gets laid off. They've done good work. They've helped teams. They've got real experience. But the market is rough, and suddenly they're out of a job. They're scrolling LinkedIn, stressed, and they start seeing the same posts over and over:

> "Tired of being an underpaid Scrum Master?"
>
> "I'll show you how to become a six-figure Agile coach."

"The industry is changing. Don't get left behind."

The message is always some version of: you're not the problem, your role is. You need to reinvent yourself with my model. They're vulnerable, scared, and looking for stability, so they book a call. On the call, the script goes something like this: "You're clearly talented. You're being undervalued. The market doesn't understand you, but I do. My proprietary system will position you as a premium coach." There's very little talk about the reality of their local market, their actual strengths and gaps, or the messy, long-term work of building trust and reputation. But there's plenty of talk about "high-ticket clients," "positioning yourself as an authority," and "using my framework to stand out."

Then comes the price: $2,000, $3,000, and sometimes more. For what? A "model" that's mostly rebranded basics. Another badge. A new label for their LinkedIn headline. Now they're under pressure. They spent real money, and they need a return. So, what do they do? They start selling the same thing forward:

"I'll help you become a six-figure coach."

"Use my method to stand out in the Agile market."

And just like that, they've been pulled into a system where the real product isn't better coaching. The real product is more people selling the product. That's not community. That's a pyramid with better branding. Meanwhile, teams still aren't getting better support. Leaders still aren't getting honest advice. The signal-to-noise ratio in the industry gets worse. And every time one of these stories ends badly, someone else walks away thinking, "Agile is a scam," or "Coaches are a scam." But the problem wasn't agility. It was the playbook.

The High-Horse Expert

There's another pattern that doesn't look as dramatic as pyramid schemes, but it does just as much damage: the high-horse expert. You've probably seen it. Someone shares a real problem or question: "We're struggling with X in our context, has anyone tried…?" Instead of curiosity from the "expert," the response quotes a framework, a guide, or a guru, but never asks even a single question about what's really going on. These folks aren't always abusive. They're not necessarily swearing at people or calling them idiots. But the vibe is clear: if you were serious about this, you'd already know. Let me correct you. High-horse experts tend to:

- Talk at people, not with them.
- Assume you're doing it wrong before they try to understand your constraints.
- Use the framework or the book as a weapon, not a tool.
- Signal how "pure" their practice is instead of helping you move from where you are.

The impact is quieter than a full-blown troll, but it's similar. Coaches and team members stop sharing experiments because they're tired of being corrected in public. Leaders who were already nervous about Agile see the pile-ons and think, "Yeah… we're not stepping into that." Nuance disappears. Fear goes up. This isn't "protecting real Agile." It's ego in a nicer outfit. Behavior-first agility applies here, too. If your online behavior shuts people down, shames them, or makes it harder for them to learn in public, I don't care how many frameworks you can quote; you're not helping.

If you want to be a grown-up in this space:

- Ask at least one question before you correct someone.
- Assume there's context you don't see.
- Offer options, not purity tests.
- Remember there's a human being on the other side of the screen, probably doing their best in a messy situation.

You can be direct. You can disagree. You can call out real harm. But if you need a high horse to feel relevant, it's worth reflecting on whether that's actually serving the community.

The Ethics of Influencers

Social media is flooded with influencers, gurus, and branded solutions, so ethics isn't optional. Ethical influence means being transparent, honest, and respectful of context.[2] It means acknowledging limits, trade-offs, and uncertainty instead of selling certainty and quick fixes. It means putting people and outcomes above hype and personal gain. Unethical influence does the opposite. It thrives on fear, identity, and dependency. It hides behind secrecy and IP.[3] It oversimplifies complex problems into catchy slogans and then blames the buyer when the promises don't pan out. Ethics matters here because leadership and change aren't products you can package and sell like software.[4] They're human, messy, and built on trust. When ethics are missing, trust erodes, cynicism grows, and real progress stalls.

Ethical influence tends to build trust by:

- Sharing evidence and admitting where the approach might not fit.
- Encouraging experimentation and learning, not blind adherence.
- Respecting the unique context of each organization and team.
- Owning mistakes and learning openly.
- Avoiding manipulative marketing tactics that prey on fear or status.

For Agile coaches, these principles align closely with the Agile Coaches Code of Ethical Conduct, which calls for respect, honesty, continuous learning, and the protection of dignity and autonomy of the people we serve.[5] For leaders and coaches, applying an ethics lens isn't optional; it's a responsibility. It means asking tough questions, not just about what works but also about how it's sold and why.

2 See Paul Gibbons, *The Science of Organizational Change* (2023), on the importance of evidence-based approaches and the risk of pseudoscience in business.

3 Ethical influence principles adapted from business ethics literature emphasizing transparency, accountability, and respect for context (e.g., Treviño and Nelson, *Managing Business Ethics*, 2016).

4 The dangers of manipulative marketing and quick-fix promises in consulting are widely documented in organizational change critiques (Kotter, *Leading Change*, 1996; Beer and Nohria, *Breaking the Code of Change*, 2000).

5 Agile Alliance. https://agilealliance.org/agile-coaching-code-of-ethical-conduct/.

Here's a quick ethics checklist you can use when evaluating any influencer, coach, or program:

- Do they provide evidence, or just anecdotes?
- Do they admit limits and trade-offs?
- Do they encourage you to experiment and adapt, or demand you follow a script?
- Do they blame clients when things don't work?
- Are they transparent about incentives, affiliations, and what they're selling?

Ethics isn't just about morality. It's about effectiveness. Without it, even the best ideas fail to take root. Applying an ethics lens helps leaders cut through the noise and make decisions that actually serve their people and purpose.

Leaders: How Not to Get Fooled

If you're a leader, founder, executive, vice president, director, or owner, you're a prime target for gurus and influencers. You control the budget. You're under pressure to transform. You want results fast. That combination makes big promises and slick decks feel tempting. So here are questions to ask any consultant, coach, or influencer before you hire them.

1. **"Tell me about a time your approach didn't work. What did you learn?"**

   If they can't give you a clear, honest example, they're either inexperienced or they're not being truthful. Neither is good.
2. **"How do you adapt your approach when the context doesn't fit?"**

   Listen for real examples of adaptation, not "we follow the framework harder." If their answer is basically "you just need to do my model correctly," then that's a red flag.

3. **"What do you need from us, as leaders, for this to work?"**
   If they don't need anything from you, be worried. Real change requires leadership behavior change, time, support, and hard decisions. If they act like they can "fix the teams" while you keep doing everything the same, they're selling fantasy.
4. **"How will we know this is working beyond your success metrics?"**
   Ask for concrete, observable changes, a mix of leading and lagging indicators, and how they'll measure impact with you, not just for their slide deck. If all they have are vanity metrics and vague "engagement scores," then be cautious.
5. **"What happens when you leave? How do we sustain this?"**
   Good partners build internal capability, make themselves less necessary over time, and have a plan for handover and sustainability. Bad ones create dependency, keep everything locked inside their model and language, and make you feel like you'll fall apart without them.

If they can't answer these questions clearly, or they get defensive, you've learned something important before you've signed anything.

Rebranding Versus Real Innovation

Not everything that looks new is actually new. In the Agile space, we tend to lump two very different things together: rebranding and real innovation. Rebranding is when someone takes existing ideas, psychological safety, systems thinking, servant leadership, or basic metrics, gives them a catchy new label or trademark, wraps them in a shiny diagram, and sells them as a "proprietary model." You've probably seen versions of this. And just to be clear, these are fictional examples, not actual trademarks at the time of writing:

- Psychological safety → "Fearless Flow™"
- Retrospectives → "Momentum Reflection Cycle™"
- Basic ROI thinking → "Value Velocity Matrix™"

Same concepts but new packaging and higher pricing. There's nothing wrong with explaining ideas in a new way to make them clearer. That can be genuinely helpful. The problem is when people pretend they invented the idea, hide the sources, or use the new label to gatekeep and upsell. That's not innovation; that's marketing. Real innovation looks different. It brings a genuinely new insight, pattern, or connection, or it finds a meaningfully better way to apply existing ideas in practice. It's being honest about what's new versus what's built on other people's work. And it makes things easier to use, not harder to access. Real innovation usually sounds more like:

- "Here's how we combined X and Y in this context, and what we learned."
- "We took this research and applied it this way—here's what surprised us."
- "This is heavily inspired by A, B, and C. Here's what we changed."

It's grounded. It's transparent. It's humble. If someone can't tell you what their work is built on, what's different, and where it doesn't fit, you're probably not looking at innovation. You're looking at a rebrand.

If You're a Coach, Read This

If you're a coach, Scrum Master, or internal change agent reading this, then there's a good chance that you're not the problem. You're probably trying to do right by your teams. You're frustrated by leadership behavior you can't control. You're watching the influencer circus and wondering if you're behind. Let me be clear: you need to keep learning. You should invest in your growth. But you don't need to reinvent yourself every six months with a new label, buy every high-ticket program that shows up in your feed, or believe the only way to be legit is to join someone's branded system. You need to be vigilant about what you learn, from whom, and what you pay for. Here's what you need:

- **Real practice with real teams.**

  Books, courses, and certifications can help, but they don't replace time in the trenches. You learn by trying things, watching what happens, and adjusting.

- **Continuous learning with discernment.**

  Read widely, take courses, go to conferences, and don't just stick to Agile; learn from psychology, systems thinking, leadership, communication, and beyond. But keep asking: *Is this grounded in something real? Can I see how this helps my context? Is this worth the time and money right now?*

- **Honest feedback.**

  Get feedback from teams, peers, and leaders. Not just "great workshop." Real feedback on what changed, or didn't, because of your work.

- **The courage to say, "I don't know."**

  You're not a vending machine for answers. You're a partner in figuring things out. Admitting what you don't know is part of being a serious professional.

- **The humility to credit other people's work.**

  You don't have to invent a model to be valuable. You can use good ideas well and be honest about where they came from.

- **A clear sense of your own ethos.**

  Who are you as a coach? What do you stand for? What won't you do, even if it pays well? That's your anchor when the market gets noisy.

And here's what you don't need:

- To chase every new acronym
- To buy into a secret system just because it's expensive
- To collect badges as proof you're worth listening to

Certifications, courses, and programs can be great if they deepen your understanding, sharpen your practice, give you tools you can use, and align with your values, not just your fear. Your value isn't your brand, your title, or your badge count. It's your ability to help people see more clearly, work more effectively, and change more safely. Build your personal brand if you want. Just build it on results, not hype. If you keep learning with intention, practicing with honesty, and keeping your ethos intact, you'll outlast the noise, and you won't need snake oil to feel confident in your craft.

We've talked about frameworks, influencers, and the whole circus that's grown up around Agile. But there's another layer that quietly poisons everything: the language we use. You can have decent intentions and still cause real damage if you hide behind jargon, slogans, and feel-good phrases that don't mean anything in practice. Words like "empowerment," "transformation," "fail fast," and "Agile mindset" get thrown around so much they become background noise, or, worse, weapons. In Chapter 4, we're going to walk straight into that mess: buzzwords, jargon, how language gets weaponized, and how to start calling it out, so you and your teams can understand each other and do real work again.

CHAPTER 4

BUZZWORDS AND WORD GAMES

A Guided Tour of the Worst Jargon

Buzzwords. We all know them. We work in an industry where you can say a full sentence and nobody really knows what you mean. We've built a way of working where words sound impressive, everyone nods along, and almost nobody can explain what any of it looks like in practice. Even some of the Agile Manifesto's authors have said it: Agile has become a buzzword.[1] That frustrates the people who actually believe in agility, and who've seen what it can do when it's done well.

The problem with buzzwords isn't just that they're annoying. It's that they turn into word games (see Figure 4.1). They let people sound decisive without committing to anything. They let leaders signal change without making one. They let teams argue over language rather than fix the work.

So in this chapter, we're going to slow down and walk through a few of the greatest hits: the phrases that sound good, get weaponized, and quietly wreck any chance of real agility if you don't call them out.

We Need to Be More Agile

Of course, we need to be more effective, more efficient, and better at working together. Everyone already knows that. Saying "we need to be more Agile" without anything concrete to back it up is just stating the obvious with extra syllables.

1 Dave Thomas, one of the original signatories of the Agile Manifesto, has repeatedly argued that "Agile" has become a meaningless buzzword and that we should focus instead on the underlying idea of "agility." See his talk "Agile Is Dead (Long Live Agility)" at GOTO 2015 (available on YouTube: https://www.youtube.com/watch?v=R4AfM2lSqB4).

BUZZWORD TYPES: THE 4 WORD GAMES

"If you can't tie it to behavior, it's noise."

VIBES

Definition: Vague concepts that feel good but lack clarity.

Tell: Sounds positive, but means nothing concrete.

SHIELDS

Definition: Used to deflect criticism or responsibility.

Tell: Evasive language that prevents accountability.

WEAPONS

Definition: Aggressive terms used to dominate or attack.

Tell: Adversarial language that creates conflict.

STICKERS

Definition: Superficial labels attached to things or people.

Tell: Quick categorization without depth or context.

"Ask: What changes on Monday?"

Figure 4.1 | Types of word games

> What people say is: "We need to be more Agile."
>
> What they often mean is: "We're not happy with how things are going, but we don't know what to change, and we don't want to admit that out loud."

That's why "be more Agile" is such a convenient phrase. It sounds decisive, but it doesn't commit you to anything. It lets you signal a change without actually making one. If you can't follow the sentence with something specific, it's just noise. It also creates a false alignment because everyone hears the same phrase and quietly fills it with their own definition. So don't argue about the word. Ask for clarity. More Agile at what, exactly? Speed, quality, decision-making, learning, or customer feedback? What are we doing today that we're willing to stop doing? What behavior needs to change first? Leaders, teams, or both? And how will we measure progress without just "more meetings happened?" Agility isn't a vibe. It's observable.

"Be more Agile" only starts to mean something when it turns into visible changes: fewer escalations and clearer ownership in decision-making; fewer urgent interruptions and better trade-offs in priorities; smaller batches and less multitasking in work in progress; shorter feedback loops and more real experiments in learning; and enough safety that people can raise bad news early without getting punished for it. Until then, "we need to be more Agile" is just a nice-sounding way to avoid saying: "We don't actually know what we're willing to change yet."

Fail Fast

"Fail fast" sounds edgy and modern. It makes people feel bold, like they're building the kind of culture where learning happens quickly, and nobody gets punished for trying. And the idea underneath it is solid. The sooner you find a problem, the cheaper and safer it is to fix. That's not new. That's basic quality assurance and risk management, and it's been true for decades, long before Agile became a brand.[2] Fast feedback loops matter because they keep small problems small. They let you adjust course before you've sunk weeks of effort into the wrong thing. Real experimentation has boundaries: a clear hypothesis, a small blast radius, and a willingness to stop when the evidence says stop. But "fail fast" doesn't always mean "learn fast." A lot of the time, it's used as a slogan to excuse sloppy work.

---

2 The idea that catching defects earlier is cheaper and less risky than catching them later has been well established in software engineering and systems development for decades. See, for example, Barry Boehm's work on the "cost of change" curve and defect reduction, such as "Software Defect Reduction Top 10 List" (IEEE Computer, 1999), and related analyses summarized by the Project Management Institute's Disciplined Agile guidance on the cost of change.

What people say is: "We fail fast."

What they often mean is: "We rush, we break things, and we don't always clean up properly, but it sounds cool when we say it."

That isn't learning. That's reckless. Failure without reflection is just damage. Failure without safety is trauma. Failure without change is waste. If you're not making space to understand why something failed, decide what you'll do differently next time, and protect your people while you experiment, then you're not failing fast. You're just failing, while burning time, money, and morale. Real "fail fast" has discipline. It looks like small bets, tight feedback, honest learning, and clean follow-through. It's not "break things and move on." It's "try something, learn something, and get better, without sacrificing the humans doing the work." So the next time you hear "fail fast," don't argue about the phrase. Ask the only question that matters: are we actually learning and improving, or are we just breaking things and hoping for the best?

We're Empowering Teams

This one sounds fantastic. If it were true everywhere, it wouldn't be a buzzword. In some places, it's real. Teams have autonomy. They can push back. They can say "no." They're trusted to make decisions, and they have the psychological safety to speak up when something feels off.[3] They're treated like adults. But in many organizations,

3 Amy C. Edmondson's ground-breaking research introduced the concept of "psychological safety," a shared belief that a team is safe for interpersonal risk-taking, which is critical for learning, innovation, and effective decision-making. Her 1999 paper, "Psychological Safety and Learning Behavior in Work Teams" (*Administrative Science Quarterly*, Vol. 44, No. 2), showed that teams with higher psychological safety were more likely to speak up, admit mistakes, and learn from one another, leading to better performance. Edmondson further explores these ideas in her book *The Fearless Organization: Creating Psychological Safety in the Workplace for Learning, Innovation, and Growth* (Wiley, 2018), which provides practical strategies and real-world examples of empowered, high-performing teams.

"empowering teams" is just a word game. It's empowerment on paper, control in practice.

It shows up in small, almost comical ways. Teams get freedom over the Jira workflow, but the real decisions still come from above. They're told they own outcomes, but scope, deadlines, and budget are still handed down like commandments. They're accountable, but not actually authorized. That gap between accountability and authority is where so much so-called empowerment dies.

What people say is: "We're empowering teams."

What they often mean is: "We've added empowerment language, but we haven't changed how power works here."

Because real empowerment isn't a vibe. It's decision rights. It's the ability to raise bad news early without getting punished. It's the ability to say, "This won't work," and have that taken seriously. It's the authority to shape how the work gets done, not just the responsibility to deliver it. If none of that has shifted, you haven't empowered anyone. You've just rebranded control. So when you hear "we're empowering teams," don't nod. Ask for something observable: who can say no here, and what happens to them when they do?

We're Outcome-Driven

This one is everywhere now. Sometimes it's true. But a lot of the time, it's just a new label on an old system. Here's what outcome-driven often looks like in the wild. There's a fixed budget that can't move. Teams are told precisely what to build. Success gets measured by output, how much we shipped, rather than outcome: what actually changed for customers or the business.[4]

4 The distinction between outputs (things produced) and outcomes (the effect or value of those things) is well established in product and Agile literature. See, for example, Aditi Agarwal, "Agile Metrics: Output vs Outcome—Measure What Matters" (2021), and Product School's "Outputs vs. Outcomes: A Costly Confusion," both of which highlight how many organizations claim to be outcome-focused while still rewarding teams primarily for output.

I've worked with, and for, organizations where year-end bonuses were tied to specific metrics, and every single one of those metrics was about output: features delivered, projects completed, hours billed. Not once in over 30 years have I seen a bonus plan tied primarily to outcomes such as customer behavior change, reduced risk, improved user quality of life, or long-term sustainability. Organizations reveal what they truly value through what they fund, reward, and protect when trade-offs get uncomfortable.

Are there genuinely outcome-driven organizations? Yes. They exist. But for most, "outcome-driven" is a sticker slapped on top of an output-driven machine.

> What people say is: "We're outcome-driven."
>
> What they often mean is: "We still reward output, but we've updated the language."

If compensation, recognition, and decision-making are all tied to output, you're not outcome-driven. You're output-driven with better marketing. Being truly outcome-driven means leaders are willing to change what they fund, what they stop, and what they reward based on whether anything meaningful actually changed for customers, staff, or the business, not just whether more stuff got shipped. So when you hear "we're outcome-driven," don't argue. Ask the question that reveals the truth: what gets rewarded here, outputs or outcomes?

We Need an Agile Mindset

"Agile mindset" might be one of the vaguest phrases we throw around. On the surface, it sounds profound: we don't just do Agile, we are Agile. We have an Agile mindset. It shows up in job postings, slide decks, and coaching pitches like it's some mystical state you unlock after enough stand-ups. And the idea underneath it isn't useless. It points to real, healthy behaviors: being open to learning, being willing to change your mind, valuing collaboration over ego, and focusing on outcomes over rituals. Those are good things. The problem is how the phrase gets used.

> What people say is: "We need an Agile mindset."
>
> What they often mean is: "We want you to accept more change, more work, and more ambiguity, with less pushback."

In some places, "Agile mindset" becomes a blunt instrument. If you raise a risk, you're not being Agile. If you question a decision, you don't have the right mindset. If you ask for clarity or limits, you're resisting change. That's not a mindset issue. That's pressure and control dressed up as coaching. Mindset, in the research sense, is about how people interpret challenge, effort, and feedback.[5] People quickly learn which kinds of honesty are safe—and which are punished. But in real organizations, the environment does most of the shaping. What gets rewarded? What gets punished? What do leaders do when someone tells the truth? You don't create an Agile mindset by telling people to think differently. You create it by changing the conditions in which people work. That means changing how decisions are made. Making it safe to surface bad news early. Rewarding learning and honesty, not just heroics. Removing punishments for speaking up or slowing down to do the right thing. If you're not willing to change structures, incentives, and leadership behavior, "Agile mindset" becomes a nicer way of saying: please adapt to our dysfunction and smile while you do it.

If you want people to think and act differently, start with what you tolerate, what you reward, and how you respond when things go wrong. The mindset will follow the environment, not the poster.

---

5 Carol S. Dweck's work on mindsets distinguishes between "fixed" and "growth" mindsets and how people view ability, effort, and learning. See *Mindset: The New Psychology of Success* (Random House, 2006). Applying this to Agile, a so-called "Agile mindset" would require environments that genuinely support experimentation, learning, and honest feedback, not just slogans about being flexible.

We Have Psychological Safety Here

Psychological safety went from an obscure academic term to a corporate buzzword in record time. On the surface, it sounds great: people feel safe to speak up here. It shows up in culture decks, values posters, and leadership keynotes. Sometimes it even shows up as a survey score that leaders brag about. And the idea itself matters. Psychological safety, in the research sense, is about whether people feel safe taking interpersonal risks at work: asking questions, admitting mistakes, challenging ideas, and raising concerns.[6] Teams with higher psychological safety learn faster, adapt better, and generally perform better. But once it became a buzzword, something predictable happened.

What people say is: "We have psychological safety here."

What they often mean is: "We did a survey once, the number looked okay, and nobody is yelling in meetings, so we assume people feel safe."

In many organizations, psychological safety is reduced to "we're nice to each other" or "we don't shout." Meanwhile, people still get punished, subtly or directly, for raising uncomfortable truths. Leaders still shut down dissent with "we don't have time for that right now." Teams still learn to keep quiet about risks until it's too late. Safety isn't measured by how calm the room feels when things are easy; it shows up in what happens when the truth is inconvenient. That's not psychological safety. That's politeness with consequences.

Real psychological safety is messier in practice. It looks like someone being able to say, "I don't think this will work," without getting labeled negative. It looks like bad news travels fast and is treated as useful information rather than disloyalty. It looks like

6 Amy C. Edmondson defines psychological safety as "a shared belief that the team is safe for interpersonal risk-taking." See "Psychological safety and Learning Behavior in Work Teams," *Administrative Science Quarterly* 44, no. 2 (1999): 350–383, and *The Fearless Organization* (Wiley, 2018), which link psychological safety to speaking up, learning, and performance.

leaders admitting when they're wrong and modeling learning in public. It looks like teams being able to talk about mistakes without the conversation turning into a witch hunt.

If you want to know whether psychological safety is real, don't look at the posters. Don't rely only on the survey score. Look at what happens when someone challenges a senior leader in a meeting. Look at what happens when a release goes badly and people are tired and stressed. Look at what happens when a team says, "We need to slow down," or "We can't take on more work." If the response is curiosity, problem-solving, and a sense of shared responsibility, you might have psychological safety. If the reaction is blame, defensiveness, or quiet retaliation, you have a buzzword. You don't create psychological safety by declaring it. You create it by changing how you respond when people take risks with the truth.

Why This Matters

These buzzwords aren't harmless. They create the illusion of alignment when there isn't any. They let leaders sound progressive without changing anything. They leave teams confused, cynical, and tired. And over time, they train people to stop listening because what is said and what is meant are often miles apart.

When language gets fuzzy, accountability gets fuzzy with it, because nobody can pin down what was actually promised. And we've only mentioned a few. We could easily spend three more chapters unpacking the other favorites: high-performing teams, servant leadership, continuous improvement, customer-centric, digital transformation, best practices, center of excellence, self-organizing… and so on. Each one has a useful idea buried inside it. And each one gets abused, watered down, or weaponized until it's mostly noise.

If we want meaningful progress, we need to stop hiding behind language and start saying what we actually mean. We need to be specific about what will change in behavior, structure, and outcomes. That's how we move from talking about change to actually making it.

How Language Gets Weaponized

Buzzwords aren't just annoying. They get weaponized. They become shields that leaders hide behind and shortcuts people use to shut down honest conversation. Once a buzzword becomes a shield, it stops helping people think and starts helping people avoid responsibility.

Take "we're Agile," for example. I've watched leaders use it to deflect criticism. Someone raises a real concern—sustainability, lack of clarity, failure to involve the right people—and the leader fires back: "We're Agile. We adapt. This is just how it is now." Translation: stop questioning this; the decision's already made. Agile becomes a shield rather than a commitment.

Then there's the classic: You're just resisting change. Raise a valid concern, maybe the framework doesn't fit your context, or people are burning out, and you'll hear, "You're not being Agile," or "You don't have the right mindset." It's the same move as defending a badge: *I'm certified in X. The guide says Y.* If you question it, you must be the problem. Instead of listening, exploring the context, or admitting that the approach might not fit, they label you resistant, so they don't have to engage. That's not leadership. It's avoidance with extra syllables.

And then there's "We're empowering teams." The message from the top is all empowerment, but senior execs still decide what goes out and when. Teams are told what to build, how to build it, and by when. Pushback gets framed as not being a team player. I've seen teams working hard to build great software, only to have leaders override priorities, force pet features, and ignore feedback from the people closest to the work and the customer, while still insisting, with a straight face, "We're empowering you." If you can't say "no," can't influence priorities, and can't push back on unrealistic timelines, you're not empowered. You're just being managed with nicer language.

How Jargon Confuses and Divides

Jargon doesn't just get weaponized. It also confuses the hell out of people. Here's how it usually plays out. People nod along in meetings while quietly thinking, *I don't know what that means.* Nobody wants to look stupid, so they stay quiet. Then the same word gets used by five different people to mean five different things. Everyone assumes they're aligned because they share vocabulary, but they aren't. And this is where it gets sneaky. Cognitive biases kick in. We hear a familiar word and assume we know what the other person means, so we fill in the gaps with our own definition. Then we wonder why everything feels off. Shared language can create the feeling of shared understanding long before real understanding exists. This is how you end up with teams thinking one thing, leaders thinking another, and everyone saying "we're aligned," only to be shocked when nothing works as expected.

It's not just annoying. It's harmful. Jargon hides real problems. It silences valid concerns. It creates fake alignment. And it makes it harder to fix anything, because people aren't even talking about the same thing. And yes, this book uses labels too: *behavior-first*, *ethos-first*, *framework performance*. I'm not pretending I'm above language. The difference is that I'm not using those phrases as shortcuts or shields. I'm using them as handles for things you can actually see and test in the real world. If behavior-first doesn't show up as observable changes—how decisions get made, how conflict gets handled, how bad news gets treated; then you should call me out, too. We need to move away from jargon as cover. That means dropping buzzwords as shields, stopping the use of labels to shut people down, and saying what we mean in plain language. Until we do that, all the frameworks, coaches, and tools in the world won't save us.

How to Call It Out and Cut Through It All

The good news is you can call out the garbage buzzwords without being aggressive. You don't need to flip a table. You don't need to embarrass anyone. Most of the time, you just need to slow the conversation down long enough for the words to touch reality. Buzzwords survive because they stay abstract (see Figure 4.2).

So the move is simple: ask questions that force the speaker and the room to connect the language to something observable. The first question is the most useful one you'll ever learn in this space:

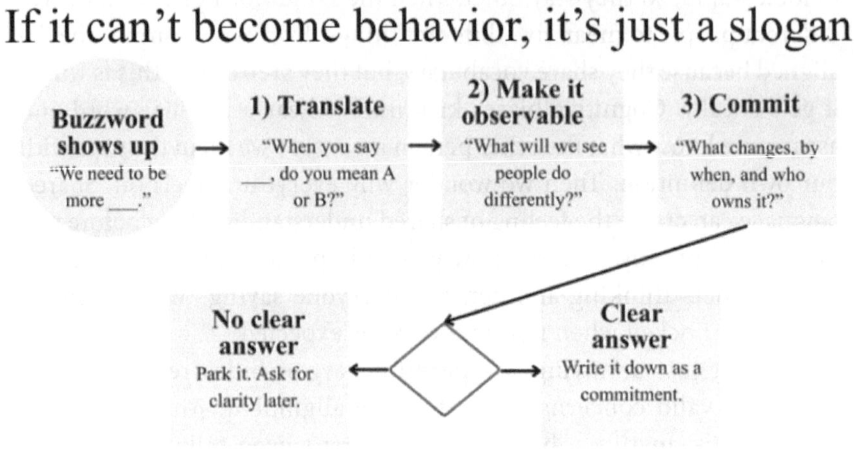

Figure 4.2 | Buzzword path

"What does that mean in practice?"

When you hear "we need more collaboration," "we're empowering teams," or "we're outcome-driven," ask what it looks like day to day. Not as a trap; as a translation request. If someone says, "You all need to collaborate more," you can respond: "Can you help us understand what that looks like in practice? What would we be doing differently if we were collaborating the way you're imagining?" Often, the person saying it hasn't thought it through. They're repeating something they've heard. This question helps them think more clearly, and it turns a vague demand into a real conversation.

The second question cuts through abstraction even faster:

"Can you give a concrete example?"

If someone says, "We're outcome-driven," try: "Can you give a concrete example of a decision we'd make differently because we're outcome-driven?" Or, if you want to go one level deeper: "How does our funding or bonus structure reflect that?" If they can't give a real example, you've learned something important: it might just be words.

And the third question is the one that turns slogans into commitments:

"What will we see that's different if that's true?"

When someone says "we're empowering teams," "we're doing a transformation," or "we're becoming more Agile," ask: "What will we see that's different if that's true?" and "How will we know, in three to six months, that this is actually happening?" Now you're talking about behaviors, decisions, structures, and outcomes, not just slogans.

These questions aren't about being difficult. They're about being honest. They help teams and leaders stop playing word games and start making real choices. They also make it harder for vague language to hide vague thinking.

Why These Questions Work

A lot of buzzwords and weaponized language don't come from evil intent. They stem from cognitive biases, habits, fear, and the very human desire to sound as if we know what we're doing. Jargon becomes a shield. It helps people avoid tough conversations. It masks uncertainty. It lets us keep moving without admitting we're not sure what moving even means. That's why questions like "What does that mean in practice?", "Can you give a concrete example?", and "What will we see that's different if that's true?" are so effective. They force everyone, including us, to slow down. They challenge assumptions without starting a fight. And they turn vague language into real things you can actually act on. You're not attacking anyone by asking these questions. You're inviting clarity. You're encouraging accountability. You're creating the conditions for real progress.

Translating Noise into Plain Language

We need to get better at translating jargon into language humans can understand and act upon. That means dropping empty slogans that sound good but don't guide behavior. It means saying exactly what we mean, with no room for guesswork. It means being specific about what will change, who will do it, and how we'll know it worked. The goal isn't to make language sound simpler for the sake of it. The goal is to make expectations, trade-offs, and behavior visible.

If you're a leader or a coach, this starts with you. You can model this in small, practical ways. Take a breath before you speak. Ask yourself: would a new team member understand this? Replace slogans with specifics: what will change, who's involved, and how we'll know it's working. You don't have to be perfect. You do have to be intentional. When you catch yourself using a buzzword, pause and translate it. When someone else uses one, ask a curious question, not a hostile one. When teams look confused, assume the language is the problem—not their intelligence.

Here's what that looks like. Every time you hear a buzzword, ask yourself: How would I say this in plain language, with actual behaviors attached?

- Instead of saying, "We need more collaboration," say what you're actually seeing and what you actually want: "Right now, design and development are working in silos. For the next two sprints, we want them to plan and review work together so we catch issues earlier and reduce rework."
- Instead of saying, "We're empowering teams," describe the decision rights and the support: "From now on, teams can say no to work that doesn't fit agreed priorities, and leadership will back them up. We'll review those priorities together every two weeks to stay aligned."
- Instead of saying, "We're outcome-driven," name the trade-off you're willing to make: "We're changing our bonus structure so it rewards customer impact and long-term value—not just how many features we ship or hours we log."
- Instead of saying, "We need to be more Agile," say what will actually change: "We're going to shorten our feedback loops and make it easier to change direction when we learn something new. That means smaller chunks of work, more frequent check-ins with customers, and fewer approval layers."
- Instead of saying, "We're doing a transformation," describe the actual shift: "Over the next 12–18 months, we're going to change how we decide priorities, how we form teams, and how leaders

behave. You'll see fewer handoffs, clearer ownership, and leaders spending more time removing obstacles than assigning tasks."

Plain language is harder than buzzwords because it requires thoughtfulness, honesty, and commitment. You have to stand behind what you say. You have to be willing to be held to it. It's not complicated. It's hard because it demands clarity and courage. The pattern is simple. Take the buzzword. Translate it into specific, observable changes in how people will work. If you can't do that, you're not ready to say it. When we get better at this, we create space for honest conversations, build trust, and make change possible. If we can ask better questions, use clearer words, and stop hiding behind jargon, we give agility a fighting chance. Because at the end of the day, agility isn't about sounding smart. It's about making it easier for people to do good work together.

Underneath the Words, It's All Behavior

This is where it loops back to behavior and ethos. Buzzwords are often used to avoid behavior change. "We're Agile" instead of "We'll stop changing priorities mid-sprint." "We empower teams" instead of "Leaders will stop overriding decisions without a conversation." "We're outcome-driven" instead of "We'll stop rewarding people purely for output." When you strip away the jargon and force yourself to speak plainly, you're left with one simple question: what will we actually do differently?

Either behaviors change, or they don't. Either power shifts, or it doesn't. Either we treat people differently, or we don't. Clear language makes that visible. Jargon hides it. Cutting through the garbage isn't just about better communication. It's about having the courage to name the fundamental changes in behavior, in leadership, and in how we treat each other that agility requires.

We've walked through the graveyard, the framework circus, the influencer ecosystem, and the swamp of buzzwords. In every one of those, the pattern is the same: people hiding behind tools, titles, and language instead of changing their behavior. You can have the right framework, the best coach, and the slickest transformation branding in the world. But if leaders still cling to control, teams still don't feel safe, and decisions still get made the same old way, nothing truly changes.

So let's strip away all of that and get to the core. In Chapter 5, we're going to talk about behavior-first agility: what it looks like in real life, why behavior change beats any process or tool, and how you start building a culture where agility isn't a label or a project—it's just how people show up, day after day.

CHAPTER 5

BEHAVIOR FIRST: THE HEART OF REAL CHANGE

By now, it should be clear: agility isn't about the framework. If you're calling yourself Agile just because you use Scrum (or any other framework), I hope you've already felt the crack in that logic. Scrum, Kanban, SAFe™, XP—these are just different ways of structuring work. They can help. They can hurt. And none of them guarantees anything on their own.

We talked earlier about Agile being an adjective, not a noun. This is where that distinction starts to matter. Because once you stop treating Agile as something you buy or install, you're left with the only part that actually changes anything: how people behave when the work gets hard (see Figure 5.1).

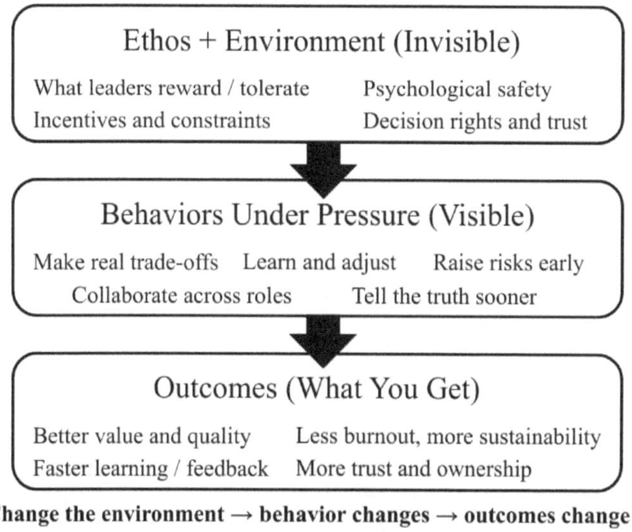

Figure 5.1 | Ethos

That's the heart of this book. Not events. Not roles. Not tools. Behavior. When priorities collide, do leaders make trade-offs, or do they push everything through and call it urgency? When a team raises a risk, do we get curious or do we punish the messenger? When a plan no longer matches reality, do we adapt, or do we double down and blame execution? When someone makes a mistake, do we learn from it, or do we hunt for someone to pin it on? Those moments are where agility lives or dies.

So in this chapter, we're going to get concrete about behavior-first agility: what it looks like in real life, why behavior change beats any process or tool, and how you start building an ethos where agility isn't a label or a project; it's just how people show up, day after day.

Behavior

Real agility is about how we react when things change, how we treat each other under pressure, how we make decisions when there's uncertainty, and how we enable ourselves and others to adapt. Those are the things your customers feel. Those are the things your teams live with every day. And those are the things that will make or break any so-called Agile transformation, no matter what's on the slide deck.[1]

One of the outcomes I want from this chapter isn't to convince you again that Agile is an adjective. It's to make that idea uncomfortably concrete, to connect the word Agile directly to the behaviors you tolerate, reward, and model, especially when things get hard. The funny thing is, we've been behaving in Agile ways our whole lives. We just didn't call it that. It was just life.

Think about how you learned to walk as a baby. You tried to stand up. You fell. You tried again. You fell again. Eventually, you stayed up a little longer. No business case. No three-month roadmap. You experimented, got feedback from gravity, and adjusted. That's agility. And it doesn't stop there. Humans adapt like this everywhere: learning to ride a bike, navigating school, adjusting to new jobs, new cities, and

1 Research on team effectiveness and organizational performance consistently points to behaviors and interaction patterns, psychological safety, clarity of goals, quality of communication, and decision-making norms as key drivers of outcomes, regardless of formal process. See, for example, J. Richard Hackman, *Leading Teams: Setting the Stage for Great Performances* (Harvard Business School Press, 2002), and Google's "Project Aristotle" findings on what makes teams effective.

new relationships. Farmers adapt to changing weather. Entrepreneurs pivot when their first idea flops. Communities improvise when disasters hit. We try, we learn, we adjust.

That basic loop—act, get feedback, adapt—is baked into how humans survive and grow, no matter where you are on the planet. Agility isn't a framework. It's a global, human default. Yet somewhere along the way, usually when we enter the world of work, we get taught something else: the boss knows everything, there's a hierarchy, and you just do your part. Those natural Agile behaviors—curiosity, experimentation, learning from failure, and helping each other out—get buried under process, fear, and hierarchy.

Behavior-first agility is about digging those behaviors back out and making them intentional again, not as a cute metaphor, but as the core of how you lead, how you design work, and how you respond when things don't go to plan.

What Behavior-First Agility Actually Means

Behavior-first agility starts with character, values, and ethos. Who you are and what you stand for drive your behavior.[2] You can bolt on as many frameworks as you like, but if the underlying ethos is fear, ego, or control, that's exactly what will show up under pressure.

At a basic human level, most of us want a few simple things. We want our work to be easier and more meaningful. We want to help others get better at what they do. And we want to cut out unnecessary friction and nonsense. Those aren't just workplace preferences; they're tied to what motivates humans in general. Maslow's hierarchy is one way of describing it: once basic needs are met, we crave belonging, growth, contribution, and purpose.[3] That's the heart of agility: making

2 Edgar H. Schein's work on organizational culture emphasizes that underlying assumptions and values (what he calls "basic underlying assumptions") drive visible behaviors far more than formal structures or slogans. See *Organizational Culture and Leadership*, 4th ed. (Jossey-Bass, 2010).

3 Abraham H. Maslow, "A Theory of Human Motivation," *Psychological Review*, Vol. 50, No. 4 (1943), pp. 370–396. Maslow's hierarchy of needs proposes that humans are motivated by a progression of needs, starting with physiological and safety requirements, then moving up to belonging, esteem, and self-actualization. In the workplace, these higher-level needs often translate into the desire for meaningful work, collaboration, growth, and a sense of purpose.

it easier to do good work together so people can move from surviving to thriving. Behavior-first agility is built on trust, real empowerment, shared responsibility, and psychological safety.

Trust is the foundation. Many organizations fail for a simple reason: people don't feel safe speaking up. They're afraid of being blamed or punished. They get judged for stepping outside the box to solve a problem. In a behavior-first environment, people can raise concerns without fear, take action to solve problems without being slapped down for not following the script, and focus on the real goal—not just their job description.

It also means breaking down the micro-silos that hide inside teams. Take a typical software team: Product Owner, developers, Scrum Master, testers. On paper, that's a team. In practice, it's often a group of people with different titles sitting near each other, defined by roles and tasks rather than real collaboration.

Compare that to a group of friends in a schoolyard: some quirky, some serious, some loud, some quiet—but they back each other up. As a group, you're stronger than you are alone. Why does that kind of behavior feel normal in life but optional at work? Behavior-first agility is about treating the team as a real team, not a collection of roles. It's about valuing the different strengths and quirks people bring, and working together toward a shared goal, not just doing your part.[4] It's less about who owns which event and more about how people show up for each other when there's a problem to solve.

Change Management Is Behavior Change

If you look at any serious work on change management, the pattern is the same: real change is behavior change.[5] Not "we implemented a new tool"

4 Patrick Lencioni's *The Five Dysfunctions of a Team* (Jossey-Bass, 2002) puts "absence of trust" at the base of a pyramid of team dysfunctions, arguing that fear of vulnerability and blame-oriented cultures undermine healthy conflict, commitment, accountability, and results. Behavior-first agility depends on addressing that foundation, not just adding new processes on top.

5 Many change frameworks emphasize that sustainable change depends on shifts in behavior and underlying assumptions, not just structures or tools. See John P. Kotter, *Leading Change* (Harvard Business School Press, 1996), and Chris Argyris and Donald A. Schön, *Organizational Learning II: Theory, Method, and Practice* (Addison-Wesley, 1996), both of which highlight that without changes in how people act and interpret their world, formal "change programs" rarely stick.

or "we rolled out a new process." Those are surface moves. Underneath, the real questions are simpler and harder: Do people behave differently now? Do leaders show up differently? Do teams interact differently? Do decisions get made differently? It's not about installing a process or tool. It's about changing the behaviors that determine how those processes and tools get used. The goal is to create conditions where people are willing to try new ways of working, understand why the change matters, and feel safe experimenting, learning, and adjusting.

Here's an early example from my own career. In the 1990s and early 2000s, test automation became the big thing in software testing. The fear was immediate and intense: "Automation is going to replace testers. We'll lose our jobs." My argument at the time was simple. Automation would handle the repetitive regression checks. That would free testers to be more strategic, more critical, and more creative. It would let them collaborate more deeply with developers, business analysts, and product people.

The tool, automation, wasn't the point. The shift was in how we saw and used testers. Stop treating them like "manual checkers." Start treating them like critical thinkers and collaborators. The same pattern shows up in Agile environments today. Processes and tools are there to support the work, but the real shift is in how people think, act, and interact. Behavior-first agility is about recognizing that the goal is not "do Scrum" or "do SAFe™." The goal is to create an ecosystem where people can work in sync, adapt together, and improve over time. You don't get that by installing a framework. You get it by changing behavior, starting with ethos and trust, and working outward from there.

Why Behavior Change Beats Any Process or Tool

So why does behavior change beat any process or tool? Because tools don't perform themselves. Frameworks don't conduct themselves. People do. You can hand someone a hammer, and they can start swinging it. They might nail a few things together. They might also dent the wood, bend the nails, or smash their thumb. They might even use the hammer to pry things apart when that's not the smartest option. If they don't understand when to use it, when not to use it, what they're trying to build, and how their actions affect the rest of the structure, it doesn't matter how good the hammer is. It might speed things up a bit, but the underlying problems remain: weak joints, flawed structures, and frustrated people. It's the same with Agile frameworks and tools.

You can train people on Scrum, teach them the roles, walk them through the events, and set up Jira boards and dashboards. You can tick every box in the playbook. But if you don't change behaviors—how leaders make decisions, how teams raise risks, how people respond to bad news—you get the same old system with new vocabulary and more meetings. Behavior is the interface between the tool and reality. If that interface is fearful, rigid, or ego-driven, no process or tool will save you.

Scrum Training, Waterfall Behavior

I saw this play out with a team in a financial institution. They were trained on Scrum. They learned the roles and responsibilities. They walked through all the events in an iteration. On paper, they were doing Scrum. In practice, they were doing Waterfall in short iterations. Nothing about behavior or governance had changed. The same heavy, Waterfall-style controls still sat on top of the team. Every iteration still required a massive 20-page test plan template. Every iteration still required formal test result documents. Approvals and sign-offs followed the same old paths.

Yes, it was a regulated environment. Yes, risk and compliance mattered. But instead of asking, "How do we adapt governance to support agility?" they bolted Scrum onto Waterfall governance and hoped the label would do the work. The result was predictable. The team went through the motions of Scrum, but the way they worked, how they decided, how they documented, and how they handled risk were still pure Waterfall. They were exhausted, and they weren't getting the benefits they were promised.

So we started again, but this time from a behavior-first angle. Over the next couple of months, we ran deeper workshops—not just on Scrum mechanics, but on how they interacted as a team. We helped them understand each other's strengths, communication styles, and blind spots. We challenged the internal silos: dev *versus* test *versus* BA *versus* the business. We worked with governance stakeholders to simplify and adapt controls so they could continue to manage risk without choking the flow. We focused on trust, shared ownership, clear goals, shorter feedback loops, and making governance support the work rather than block it.

Within three months, their productivity jumped by roughly threefold. The quality of their work improved dramatically. And the team started to feel like a real team, not a set of roles stapled together. Same people. Same domain. Same basic framework. The difference was in behavior. Silos softened. Governance adapted. That's why behavior change beats any process or tool. A great framework with old behaviors will fail. A decent framework with healthy behaviors will usually succeed. If you don't change how people think, act, and interact, no tool or process will save you.

How Do You Create a Behavior-First Culture?

So the big question is: How do you create a behavior-first culture in an Agile environment? Let's keep it simple. Start with the basics: treat people the way they'd want to be treated. That sounds obvious, but most organizations don't live it—especially under pressure.

A behavior-first culture starts with respect that shows up in real moments, not posters. It shows up in how we talk to each other when deadlines are tight and something breaks. It shows up in assuming good intent—because most people genuinely want to do a good job. And it shows up in recognizing value: people want their work to matter, not just fill a backlog. If you genuinely believe, "Everyone here wants to do something meaningful," and "Everyone here wants to contribute to something that has value," you stop treating people like resources and start treating them like partners.[6] That's the foundation.

From there, align on a real shared goal, not fake consensus. Behavior-first culture isn't just about being nice. It's about treating each other well and committing to something clear together. This is where *The Five Dysfunctions of a Team* gets it right: you don't need everyone pretending to agree. You need real commitment. "This is what we're

6 Douglas McGregor's Theory X and Theory Y highlight how leaders' core assumptions about people ("they're lazy and need control" vs. "they want to contribute and grow") shape management behavior and, ultimately, performance. See *The Human Side of Enterprise* (McGraw-Hill, 1960).

trying to achieve." "We all understand it." "We may have argued about how to get there, but now we're aligned."

And then, only then, use frameworks as helpers, not cages. Scrum, Kanban, SAFe™, XP—whatever you're using should support the behaviors you want. It should make it easier to collaborate, learn, and adapt. It should help you see problems sooner. It should never override basic respect, replace thinking, or become more important than the people using it. A behavior-first culture starts with how we treat each other. It aligns around shared, meaningful goals. And it uses frameworks and tools as supporting actors, not the main character. When you focus there first, the mechanics of being Agile get a lot easier.

Process First Versus Behavior First

Figure 5.2 shows the difference between process first and behavior first in practice. Process-first thinking sounds like compliance. The questions are usually some version of: Did we follow the framework correctly? Did

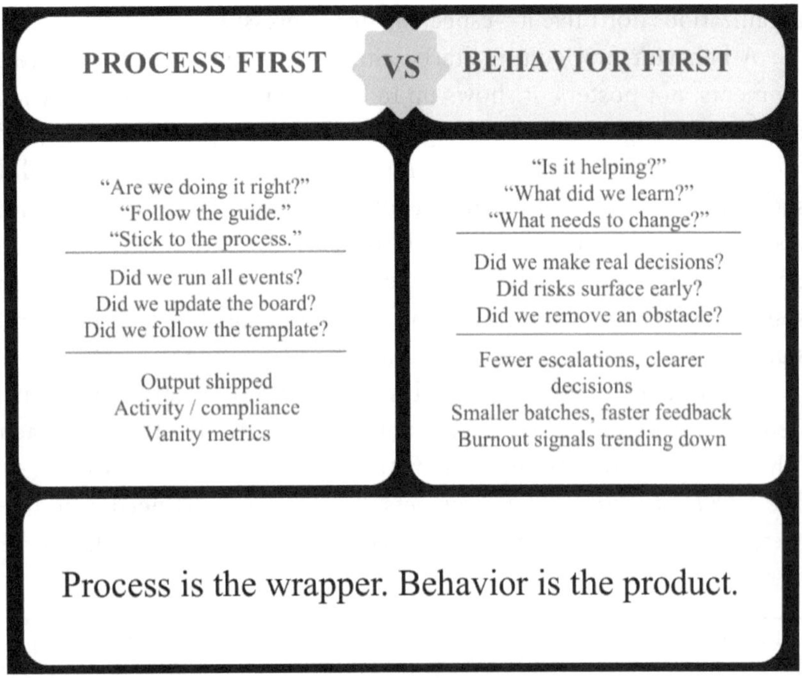

Figure 5.2 | Process versus behavior

we run all the events on the calendar? Did we fill out the templates and check the boxes? Did we update Jira? The focus isn't on outcomes. It's on whether we "did it right."

Behavior-first thinking sounds different. It asks: Did people feel safe to speak up when something was wrong? Did we learn something and adjust based on that learning? Did we make it easier to do the right thing next time? Did we remove at least one obstacle that was slowing us down? The focus is on how we show up and what has actually changed.

Here's the uncomfortable part: you can be 100% process-compliant and still be rigid, fearful, and slow. And you can be a bit messy in the process, yet genuinely Agile if the behaviors are healthy. Ideally, you want both: simple, supportive processes and strong behaviors. But if you have to choose where to start, start with behavior. That's the part that makes everything else either work or collapse.

Everyday Behaviors That Signal Real Agility

If you want to know whether agility is real in your environment, don't start with the framework. Start with the small moments. Real agility shows up when a leader can say, "I was wrong about that. Let's change course," and the team respects it instead of punishing the honesty. It shows up when someone raises a risk early, and the response is "Thank you," not "Why are you being negative?" It shows up when a team member says, "I don't know," and the next move is, "Okay, how do we find out?" instead of shame.

You see it in who gets invited to discovery and planning, and who is only pulled in at the end to sign off. You see it when a team pushes back on unrealistic scope or timelines, and leadership actually listens and adjusts. You see it after something goes wrong: Do people ask, "What did we learn?" or "Who do we blame?" You see it when someone spots a better way to do something and gets encouraged to try it, even if it's not "how we've always done it." And you see it in whether cross-functional work is normal. Not as a special workshop. Not as a quarterly alignment session. Just everyday collaboration: developers, testers, analysts, ops, and business folks talking regularly, not only at handoff points. These behaviors look small in the moment, but they add up. You don't need a slide that says, "We're Agile." You need these moments to happen repeatedly. If you start seeing these patterns, you're probably closer to real agility than a dozen framework rollouts will ever get you.

Behavior-First Self-Reflection (for Leaders and Teams)

To make behavior-first agility real, we start by holding the mirror up to ourselves. This isn't about blame or perfection. It's about honest reflection and intentional change. You can use questions like these as a leader, a coach, a team member, or a whole team.

Start with yourself. What's one behavior you're willing to change in the next month? Maybe it's interrupting less. Asking more open questions. Admitting when you don't know. Or actually listening when someone raises a concern—even when it's uncomfortable. And where do you hide behind process, jargon, or rituals instead of being honest? Maybe it's saying "we're Agile" when what you really mean is "we're changing things, and I know it's messy." Maybe it's defaulting to a framework instead of addressing the real issue.

Then zoom out to the team. What's one behavior you want to see more of? Raising issues early. Asking for help. Pairing more. Challenging unclear priorities. The kinds of behaviors that build trust and collaboration. And what's one behavior you need to stop rewarding? Heroics that burn people out. Saying yes to everything without question. Staying silent in retrospectives and then complaining later.

Finally, look at the organization. What's one behavior you say you value but don't actually reward? Experimentation. Speaking truth to power. Pushing back on bad work. These often get lip service and little real support. And what's one behavior you quietly reward that works against agility? Individual heroics over team success. Hoarding information. Hiding bad news. The small habits that undermine trust and transparency. These questions aren't just a nice exercise. They're a call to action.

Real behavior change requires leaders to pay attention to more than process. It requires paying attention to the human environment: how stress shows up, how people communicate under pressure, and whether it's safe to tell the truth early. If your people are exhausted, afraid to speak up, or constantly bracing for the next fire drill, no amount of Agile on a slide deck will fix that. You're burning humans to keep a framework alive. Behavior-first agility is about changing that reality, one behavior at a time.

The next chapter dives into the heart of it: team well-being. Why psychological safety and trust matter more than sprints or story points, how *Agile done wrong* leads to burnout and stress, and what it takes to build teams that don't just survive change, but genuinely thrive.

CHAPTER 6

TEAM WELL-BEING OVER EVERYTHING

In recent years, team well-being has finally started getting attention, not because leaders suddenly became nicer, but because the cost of ignoring it has become impossible to hide. Burnout is everywhere. Turnover is expensive. Work has gotten more complex. The pace of change is relentless. After years of remote and hybrid work, economic uncertainty, and constant disruption, many people are running on fumes. So now we're talking more openly about stress and what's happening at home as well as at work, and words like mental health, burnout, and boundaries are showing up in leadership conversations. That's a good thing.

The thing is, the best leaders have been paying attention to well-being long before it became the next big thing on LinkedIn. Much of this current wave began when work–life balance became the buzzword of choice. And if you're anything like me, when you hear "work–life balance," your eyes probably roll a bit, not because the idea is bad, but because we've been sold the promise before. It works for a bit. Then everything quietly slides back to normal.

Just like with frameworks, influencers, and buzzwords, many of us have been burned by fake versions of "we care about well-being." Pizza, yoga, and a mindfulness app on top of the same old pressure cooker don't, by themselves, create a healthy environment (see Figure 6.1).[1]

---

1 Research on burnout and engagement consistently shows that well-being is driven less by individual "self-care" tactics and more by workload, control, fairness, values alignment, and social support. See Christina Maslach and Michael P. Leiter, *The Truth About Burnout: How Organizations Cause Personal Stress and What to Do About It* (San Francisco: Jossey-Bass, 1997), and their later work, summarized in *Burnout* (Harvard Business Review Press, 2022), which argues that organizational conditions—not just individual resilience—are the primary drivers of burnout.

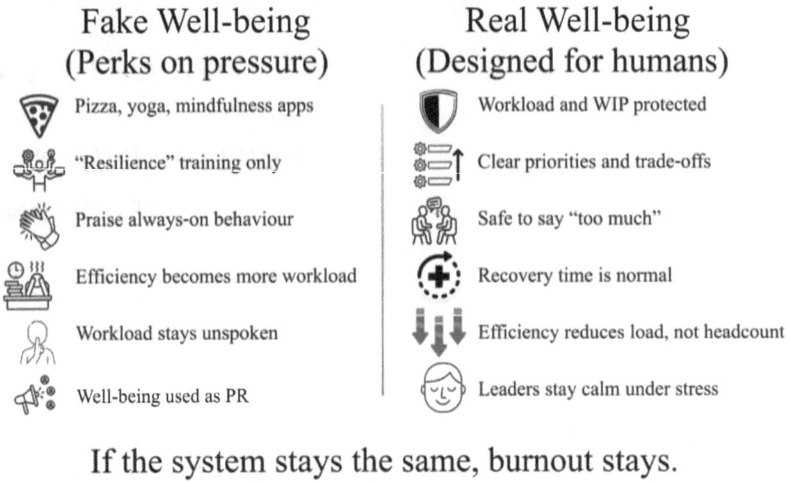

Figure 6.1 | What does well-being really mean

In the previous chapter, we talked about behavior-first agility. Well-being is where that gets real. If your behaviors don't protect people's energy, health, and safety, you're not behavior-first in any meaningful way—you're just using nicer language while people burn out.

So let's cut through it. If you want a truly Agile environment, you need people who are physically and mentally present. Not just showing up, but actually able to think, collaborate, and care. You don't get that by squeezing every drop of productivity out of them and then slapping self-care on a slide. You get it by designing work, expectations, and leadership behaviors so people don't burn out in the first place.

Psychological Safety and Trust: Sprints and Story Points

If you remember nothing else from this chapter, remember this: psychological safety and trust matter more than your sprints. You can have perfect sprint planning, beautiful boards, and clean burn-down charts. But if people are afraid to speak up, afraid to say, "I'm overloaded," or "this doesn't make sense," or "we're about to break something," you don't have an Agile team. You have a fragile one.

Psychological safety and trust are what make it possible for people to voice ideas that go against the grain, raise concerns before they become disasters, and build on each other's thinking instead of shutting it down. For example, imagine a team member says in sprint planning, "I don't think we can finish all of this without cutting corners." In a low-trust environment, the response is immediate: "We already committed," or "Don't be negative," or "Just do your best." The message is clear: stop raising concerns and fall in line. So next time, that person stays quiet. The risk doesn't go away. It just goes underground.

In a high-trust environment, the response sounds different: "Okay, what are you seeing that we need to pay attention to?" Now the team can talk honestly about capacity, trade-offs, and risk while there's still time to adjust. That's what psychological safety looks like in practice. Not comfort. Not endless agreement. The ability to tell the truth early without getting punished for it. There's a simple pattern you can hear when trust is present. It sounds like good improv and good conversation: *yes, and.*

"*Yes*, I hear you, *and* here's another angle."

"*Yes*, that's a risk, *and* here's how we might reduce it."

"*Yes*, that didn't work, *and* here's what we learned."

When there's trust, people feel safe to add, challenge, and build. That's where creativity and real problem-solving come from. And this isn't just an Agile thing. It's basic human work: how we relate to one another, how we handle stress, and how we evolve together. Agility just makes it harder to hide when it's missing. The more you shorten feedback loops and increase change, the more a lack of safety and trust shows up as anxiety, silence, and burnout rather than learning.

The AI Trap and the Hidden Cost of Efficiency

Right now, artificial intelligence (AI) is taking over repetitive work, speeding up certain tasks, and helping us draft, analyze, and automate. That can be great, if we use it wisely. But here's the trap I'm already seeing. Someone says, "AI saved us 20% of our time." A leader replies, "Great—so now we can do 20% more work." Or worse: "Maybe we can cut the team from five people to four and keep the same output." On a spreadsheet, that looks logical. In real life, it's a disaster.

Here's what happens. The remaining four people carry more of the load. Expectations quietly creep up. The saved time gets filled with more tasks, more projects, more pressure. Stress rises, burnout risk increases, and quality starts to slip.[2] And when burnout hits, it's not just that someone's a bit tired. It means people needing time off, mentally checking out, or leaving altogether. Now you've got a gap in the team, the cost and time of hiring or bringing in a contractor, and the productivity dip that always comes with onboarding. Maybe productivity climbs again eventually. It might even surpass where you were. But at what cost? Human cost. Trust cost. Cultural cost.

If you save 20% and the first thing leadership does is cut a person, you've taught the organization a brutal lesson: if we get more efficient, someone loses their job. Congratulations, you've set a match to trust, psychological safety, and people's willingness to experiment or speak up. And once people learn that lesson, they adapt in exactly the wrong way. They stop sharing ideas that could improve the system. They hold back on automation suggestions. They become careful, guarded, and politically savvy rather than open and inventive. Improvement stops feeling like contribution and starts feeling like self-sabotage.

Imagine someone on the team finds a way to use AI to take two hours of repetitive reporting off their plate every week. For the first time in a while, they feel a little relief. Maybe they can think more clearly. Maybe they can spend more time helping a teammate, catching a risk early, or just finishing the day with a bit of energy left. But if the response from leadership is, "Great, now you can take on more," or worse, "Maybe we don't need as many people," that relief disappears fast. The lesson lands hard: don't make things better if better just means more pressure. After that, people stop sharing. Not because they're lazy, but because they're protecting themselves. That's the hidden cost leaders miss. The damage isn't just immediate overload. It's the long-term message: don't make the system better if being useful makes you expendable.

---

2 Burnout research consistently links chronic workload, lack of control, and fear about job security to exhaustion and disengagement. See Christina Maslach and Michael P. Leiter, *The Truth About Burnout: How Organizations Cause Personal Stress and What to Do About It* (San Francisco: Jossey-Bass, 1997), which highlights how "doing more with less" cultures erode well-being and trust over time.

The deeper issue isn't AI. It's what leadership does with the capacity AI creates. When reclaimed time is treated as free labor, people learn that efficiency is risky. They become less likely to share improvements, suggest automation, or surface ideas that could make the system better because the pattern they've seen is simple: "If we get more efficient, expectations go up, or headcount goes down." That's not a mindset problem. It's a trust problem. And over time, it quietly undermines the very behaviors organizations say they want: learning, transparency, and continuous improvement.

AI can reduce some of the grind. It can free up time for deeper thinking, better collaboration, better customer conversations, and actual recovery. It can help people spend less time on repetitive admin and more time on judgment, creativity, and problem-solving. But that only happens if leaders make a clear, credible commitment to using efficiency gains to improve the work and protect the humans doing it, not to punish them for being successful.[3] AI won't save you from a broken culture. It just makes the cracks show up faster.

Burnout, Stress, and Agile Done Wrong

A lot of Agile discourse focuses on productivity, quality, and speed. In a healthy, behavior-first environment, those can be real byproducts. They're what happens when people have clarity, trust, and the space to learn. But when you chase those outcomes without protecting well-being, you get a very different picture. People work longer hours to hit commitments. Teams context-switch constantly because "we're Agile." Stress climbs. Quiet quitting shows up. Cynicism spreads. Turnover becomes normal. And eventually, burnout and chronic stress kill the very things you were trying to improve. Quality drops. Productivity drops. Creativity disappears.

And the damage usually doesn't arrive all at once. It shows up first in smaller signals. People stop volunteering ideas. They become more

3 Maslach and Leiter's work on burnout shows that chronic overload, lack of control, and value conflicts lead to exhaustion, cynicism, and reduced effectiveness, undermining quality and performance over time. See Christina Maslach and Michael P. Leiter, *The Truth About Burnout: How Organizations Cause Personal Stress and What to Do About It* (San Francisco: Jossey-Bass, 1997).

defensive. Conversations get shorter, flatter, and more transactional. Teams start doing whatever is safest politically instead of what is smartest operationally. Energy drops before output does. By the time the numbers look bad, the human cost has usually been building for a while. Then you pay for it twice. You pay to fix defects. You pay to replace people. You pay in lost knowledge and broken trust every time someone leaves.

That's the hidden cost of Agile done wrong. You might get a short-term bump in output. But long term, you're burning through people instead of building sustainable performance. And here's the uncomfortable truth: a behavior-first approach that ignores well-being isn't behavior-first at all. It's just control and urgency in Agile clothing.

Well-being Is the System

We'll dig into this more in the rest of the chapter, but the core is simple: well-being is infrastructure. Psychological safety and trust are the foundation. And AI and tools should reduce stress, not become an excuse to cut people and push harder. Teams thrive when the work is designed for humans. That means designing a flow so people can focus without constant thrashing. It means treating efficiency gains as a chance to improve quality, learning, and sustainability, not as a reason to squeeze harder or quietly reduce headcount. It means making it safe to say, "This is too much," or "This isn't working," and treating that as useful information, not disloyalty.

It also means recognizing that well-being is not separate from performance. It shapes performance. When people are overloaded, guarded, and running on fumes, decision-making gets worse. Collaboration gets thinner. Learning slows down. Small problems become bigger ones because no one has the capacity to address them early. What looks like a people issue is often a system issue. And it means caring as much about the humans as you do about the throughput. In a real, behavior-first Agile environment, the goal isn't just to move faster. It's to create a place where people can do good work together for a long time, without destroying themselves in the process.

Stretch, Don't Snap: Workload and Calm Leadership

One of the most important parts of team well-being is how you handle workload. There's a useful distinction between *pressure* and *stress*, as noted in *Relentless* by Tim Grover and Shari Wenk.[4] Pressure is when the goal is challenging but achievable. People are stretched a bit. There's a clear target and a clear reason. When you hit it, you feel proud and energized. Stress is when the goal is unclear or impossible, the load never lets up, you don't feel in control, and even when you hit the target, you mostly feel relief, followed immediately by dread about the next one (see Figure 6.2).

That difference matters. Healthy teams need pressure. That's where growth, learning, and pride come from. But when you cross the line into chronic stress, you're burning people, not building them. In a healthy, behavior-first environment, you use sprints and iterations to find that sweet spot. The conversation sounds more like: "This is a stretch, but we believe we can do it. If we can't, we'll talk about it early, not hide it." Not: "We committed, so just make it happen."

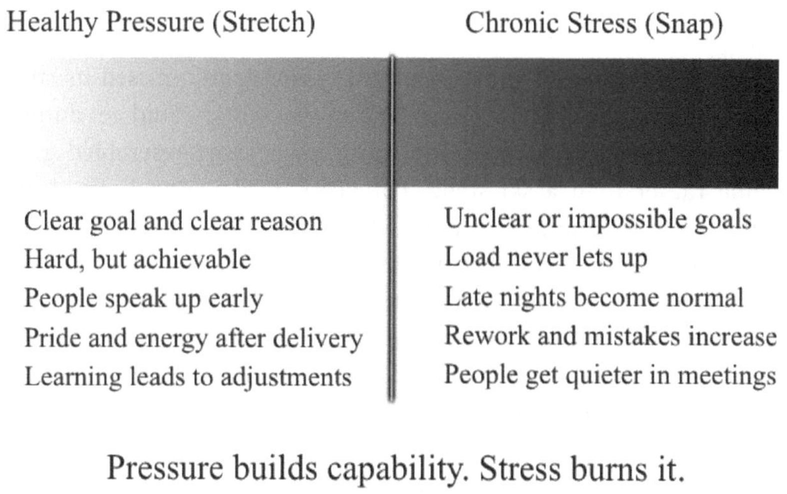

Figure 6.2 | Stretch versus snap

4 The distinction between healthy pressure and harmful stress shows up across performance and well-being research: challenge can enhance focus and growth, while chronic, uncontrollable demands lead to burnout. See, for example, Tim S. Grover and Shari Wenk in *Relentless* (Scriber, 2013), and Christina Maslach and Michael P. Leiter's *The Truth About Burnout: How Organizations Cause Personal Stress and What to Do About It* (San Francisco: Jossey-Bass, 1997).

You also pay attention to the early signals that pressure is turning into stress. People constantly working late. Jokes about burnout that stop being funny. Mistakes and rework creeping up. Tempers getting shorter. People getting quieter in meetings, not more engaged. Those aren't personality flaws. They're data points. When you see those signals, you adjust the load. Reduce the scope instead of pretending everything is fine. Resequence work so the team isn't juggling five priorities at once. Bring in help when it's genuinely needed. Say no to extra demands instead of absorbing them and hoping for the best.

The goal is to keep people in the zone where they're a little stretched, they're growing, and they feel great when they hit the goal, not the zone where they're just surviving and counting down to a vacation. And over time, you can increase that healthy pressure threshold, not by piling on more work, but by building real capacity: clearer priorities, better decision-making, fewer handoffs, stronger skills, and enough recovery that people can meet a hard moment without snapping. That's what sustainable performance looks like. It's not endless intensity, but a system that helps people handle pressure with more resilience and less damage.

I saw this play out with a team that consistently missed its sprint goals. Their intake process was so broken that work would get dumped into the sprint at any time. The burn-down chart resembled a city skyline rather than a ski slope. For years, leadership obsessed over hitting a specific number of points per sprint, and it was a hot mess. Eventually, after months of struggling and a couple of angry calls from the CTO about the need for a reset, the team finally got permission to start over. They stopped fixating on points and focused on getting the process and behaviors right: clearer intake, fewer last-minute surprises, better conversations about capacity, and more honesty about what was actually possible.

After just a few sprints, they found their groove and started delivering more value than ever. Not because they tried harder, but because they were finally working in a way that fit how humans actually get things done. Once the work stopped ambushing them mid-sprint, the pressure became manageable again, rather than constant stress.

Norms Around Availability: Remote, Hybrid, and Reality

The genie is out of the bottle with remote work. The pandemic proved something many leaders didn't want to believe: remote work can be done, value can be created, and deadlines can be met without everyone sitting in the same building five days a week.[5] Organizations demanding a full return to the office are trying to drag people back to a year-2000 way of working in a 2020s world.

That doesn't mean never see each other. One or two days a week in the office can still be valuable for connection, collaboration, whiteboard sessions, and the kinds of conversations that are often easier face to face. The point is, we can design for that intentionally instead of pretending we need butts in seats to prove people are working.[6]

In a healthy, behavior-first environment, you set explicit norms around availability. You agree on core hours when people are generally reachable. You set expectations about response times, like "we don't expect instant replies to every ping." You carve out no-meeting blocks for deep work and protect them as if they matter, because they do. You also respect time zones, home lives, and boundaries. That means not assuming everyone is available early, late, and on weekends. It means not praising people for always being on. And when you bring people together in person, you use that time for what it's actually good at: relationship building, strategy, creative work, and the kinds of conversations that benefit from being face-to-face, not just sitting in different chairs while everyone takes Zoom calls from the office.

For example, a healthy leader might say, "I don't expect you to answer messages after hours unless something is truly urgent. If I send

5 Multiple large-scale studies during and after the COVID-19 pandemic found that many knowledge-work teams maintained or even improved productivity while working remotely, with outcomes heavily influenced by management practices and clarity of expectations rather than physical presence alone. See, for example, Emma Harrington and Natalia Emanuel, "Working Remotely? Selection, Treatment, and the Market for Remote Work," NBER Working Paper No. 30862 (2022), and various summaries in *MIT Sloan Management Review* on remote/hybrid productivity.

6 Research on hybrid work has found that employees working from home two days a week were just as productive and just as likely to be promoted as fully office-based peers, while also improving retention, suggesting that intentional hybrid design can support both performance and flexibility (https://news.stanford.edu/stories/2024/06/hybrid-work-is-a-win-win-win-for-companies-workers).

something late, it's because that's when I'm catching up, not because I need you online." That one sentence does a lot. It removes guesswork, lowers pressure, and lets people know they won't be punished for setting boundaries.

This is where many organizations get it wrong. They say they support flexibility, but the hidden rules tell a different story. Messages sent late at night still carry pressure. Leaders still reward the fastest responder. People still feel they have to prove they're working because they're not physically visible. So even when the policy sounds modern, the behavior underneath it is still built on surveillance, anxiety, and performative availability. That's what makes this a norms issue, not just a logistics issue. If the unspoken expectation is "always be reachable," remote work becomes a stress amplifier. If the expectation is "do good work, communicate clearly, and use your time like an adult," remote and hybrid work can actually reduce stress and improve performance.

Remote and hybrid work, done well, can reduce stress: less commuting, more flexibility, more control over the work environment. Done poorly, they can increase stress: always-on culture, blurred boundaries, and the unspoken expectation that if you're at home, you can always take one more call. The difference isn't the location. It's the norms and behaviors around availability.

How Leaders Respond to Stress (Everyone's Watching)

One thing I've learned as a senior leader is that everyone is watching how leaders respond to stress. If a leader panics, yells, blames, and starts thrashing, changing direction every hour, the team will do the same. They'll panic. They'll hide. They'll stop telling the truth. They'll start protecting themselves instead of solving problems. If a leader acknowledges the pressure, stays calm and grounded, asks good questions, and focuses on facts and options instead of blame, the team feels safer. They bring issues up earlier. They think more clearly. They work on the problem instead of the politics.[7]

7 Research on emotional contagion and leadership shows that leaders' emotional states and reactions strongly influence team stress, decision-making, and performance. See Daniel Goleman,

In a healthy, behavior-first environment, leaders model regulated behavior under stress. They say things like, "Okay, this is a lot. Let's slow down for ten minutes and get clear on what's happening." Or, "I'm frustrated, but I'm not going to take that out on you. Let's figure this out together."

They separate urgency from panic. "This is urgent. We need to move fast." Not: "Everyone drop everything, we're all screwed, I don't care what else you're doing." And they make it safe to tell the truth. "If something's going wrong, I want to hear it early, while we can still fix it." Then they prove it by not shooting the messenger. If leaders freak out, teams will too. If leaders burn themselves out, teams will assume that's the expectation.

If you want teams that thrive, not just survive, you handle workload so people are stretched, not snapped. You set availability norms that respect humans, not just calendars. You lead in a way that makes it safe to be honest, even when things are on fire. That's team well-being over everything. If there's one thing I want you to take from this chapter, it's this: you don't get real agility by squeezing people harder and calling it "high performance." You get it by building an environment where people can think clearly, speak honestly, and stretch without snapping.

Psychological safety and trust matter more than your sprints. Workload needs to live in that healthy pressure zone, not chronic stress. Remote and hybrid work need sane norms, not 24/7 availability. Leaders need to model calm, grounded behavior under pressure because everyone takes their cue from them.

When you put well-being first, productivity, quality, and speed don't disappear. They become sustainable byproducts, not short-term gains. And that brings us to the people who either make all of this possible or quietly destroy it: leaders.

In Chapter 7, we'll dig into leadership lessons learned the hard way: what authentic leadership looks like in chaos, why courage and humility matter more than titles, and real stories of leaders who either stepped up or fell apart when it counted.

Primal Leadership: Unleashing the Power of Emotional Intelligence (Harvard Business School Press, 2002), which argues that a leader's ability to stay calm, self-aware, and emotionally regulated under pressure is a key driver of team climate and effectiveness.

CHAPTER 7

LEADERSHIP LESSONS (THE HARD WAY)

Before we get into leadership lessons, let's get clear on one thing: what is a leader? Ask most people, and you'll hear some version of "the boss," "the person who runs the show," or "the one in charge." But we need to move away from that. A boss isn't automatically a leader.

What Is a Leader?

A leader is someone who helps individuals and teams become better, not just at what they do, but at how they work together. They create conditions for other people to succeed. They make it easier to collaborate, learn, recover, and keep moving when things go sideways. Leadership is about behavior and impact, not a box on the org chart. It's also not just about authority. It's about influence, responsibility, and what your behavior makes possible for other people.[1] If people feel clearer, safer, and more capable after interacting with you, that's leadership. If they feel smaller, more confused, or more guarded, that's leadership too, just not the kind we should be aiming for.

Good leadership doesn't always look dramatic. A lot of the time, it looks ordinary: helping people think more clearly, making it safer to speak honestly, and reducing friction instead of adding to it (see Figure 7.1).

1 Many leadership scholars define leadership as a process of influence and relationship rather than a function of title alone. See Joseph C. Rost, *Leadership for the Twenty-First Century* (Westport, CT: Praeger, 1991); Peter G. Northouse, *Leadership: Theory and Practice*, 9th ed. (Thousand Oaks, CA: Sage, 2021).

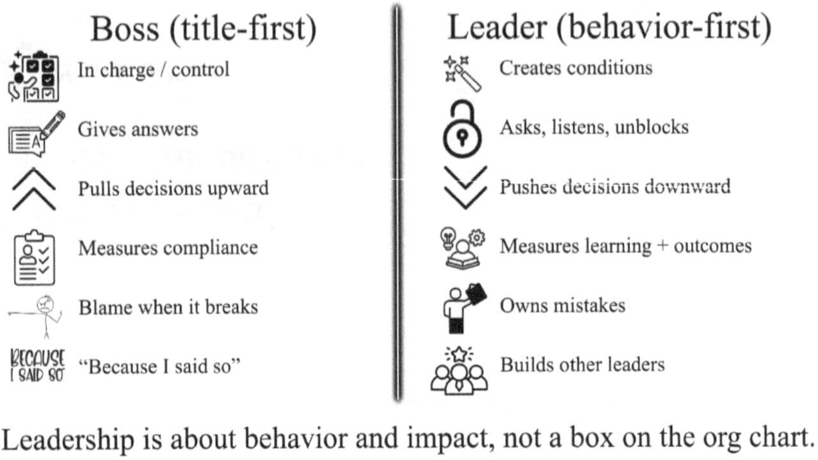

Figure 7.1 | A boss versus a leader

Here's the part a lot of people miss: if you're reading this, you're a leader. It doesn't matter if your title is Marketing Specialist, Software Developer, Accountant, Product Owner, Scrum Master, or something else entirely. You can improve how you work. You can help the people around you get better. You can influence how decisions get made, how conflict gets handled, and how truth travels. That's leadership.

And if you do have a formal leadership title—manager, director, VP, executive—it doesn't put you above anyone. It means you carry more responsibility. Your behavior has a bigger blast radius. You have more power to make things better or worse, faster, and on a larger scale. The title doesn't make you a leader. It increases the consequences of how you lead. What you create tends to spread.

If you've made it this far in the book, I hope one thing is clear: we can all make work better for the people around us. We are all leaders, in one way or another.

What Does Authentic Leadership Look Like When Things Go Sideways?

The short answer is: it shouldn't look that different from leadership when things are calm. Chaos usually means something unexpected happened, the situation is unclear, and there's pressure to respond. The temptation is to "do leadership" loudly, circle the wagons, spam everyone with emergency meetings, and drag half the organization into a panic call to get to the bottom of things. Most of the time, that reaction adds more chaos than it removes. It may feel decisive in the moment, but a lot of it is just stress wearing a leadership costume.

Authentic leadership in chaos is quieter and more deliberate (see Figure 7.2). Under pressure, weak leadership often gets louder. Strong leadership becomes clearer. It creates enough calm for people to think, enough direction for people to move, and enough trust for people to tell the truth before the problem gets worse. That doesn't mean doing nothing. It means not confusing motion with progress.

First, you observe what happened. What do we know for sure? What's still unclear? What's noise? Then you give calm, clear guidance: "Here's what we're doing in the next hour (or day). Here's who's working on what. Here's when we'll regroup." You create just enough order for

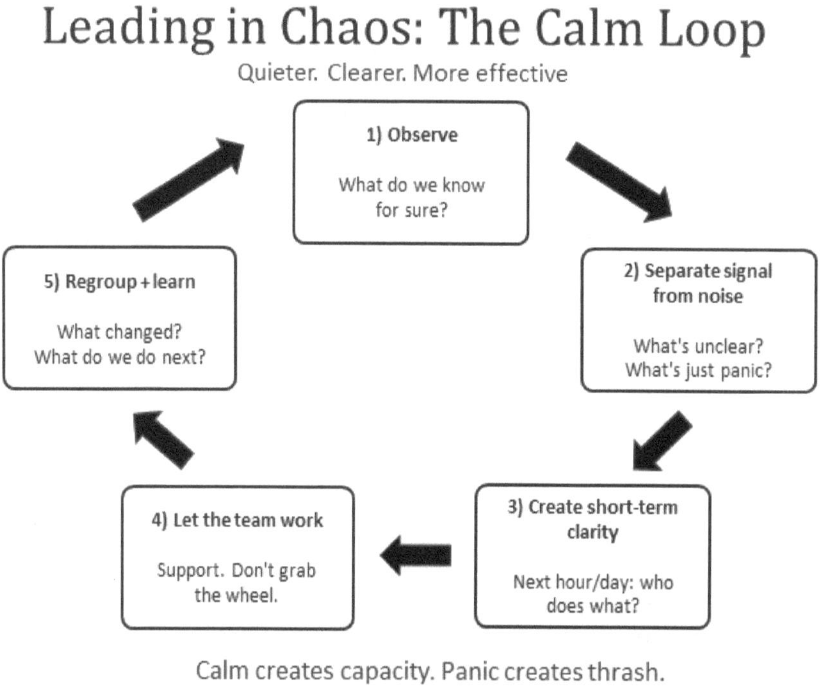

Figure 7.2

people to move without creating so much control that nobody can think. Then you let the team work the problem. If they can handle it, let them. If they're stuck, bring the right people together on purpose, not as a knee-jerk reaction. Not everyone needs to be involved in every problem. Pulling in too many people too quickly usually creates confusion, delay, and a whole new layer of noise to sort through.

Leadership style shifts with context. Sometimes you ask and empower. Sometimes you decide and direct. But the pattern stays simple: make sense of what's happening, create some order, and work toward a solution.

Here's the link back to the last chapter: if you've built trust, psychological safety, and leadership within the team, then in many chaotic situations, the team should be able to handle most of them. Leadership's job is to support, stay informed, and regulate the temperature, not to take over and make everything worse.[2]

When I was an executive, people would come into my office with problems: "We're running into this issue..." My first question was almost always, "Okay, what do you think the solution is?" Most of the time, they already had a good idea. They didn't need me to rescue them. They needed space to think, a bit of support, and permission to act. So, my response became, "Great. Go ahead and try it. Keep me posted." That's leadership. You don't grab the steering wheel every time there's a bump. You trust the people closest to the work, and you use the team's collective intelligence, which is almost always greater than yours alone.[3]

Courage, Humility, and Learning from Failure

If you want to be that kind of leader, start here: stop pretending leadership means having all the answers. Authentic leadership in a behavior-first

2 Research on emotional contagion and leadership shows that leaders' emotional states and reactions strongly influence team stress, decision-making, and performance. See Daniel Goleman, Richard Boyatzis, and Annie McKee, *Primal Leadership: Unleashing the Power of Emotional Intelligence* (Boston: Harvard Business School Press, 2002), which argues that a leader's ability to stay calm, self-aware, and emotionally regulated under pressure is a key driver of team climate and effectiveness.

3 Studies on collective intelligence suggest that groups with good communication, social sensitivity, and equal participation often outperform even their most capable individual members on complex tasks. See Anita Williams Woolley et al., "Evidence for a Collective Intelligence Factor in the Performance of Human Groups," *Science* 330, no. 6004 (2010): 686–688.

world is built on three things: courage, humility, and the ability to learn from failure.

Courage

Courage is being able to say "I don't know" and mean it. It's asking the team for help instead of pretending. It's admitting when you've made a mistake, even a big one. I've had to do this myself. As a senior executive, I made a decision about how teams would work that was pure command-and-control. It went against everything I'd been building with them over the years. External pressure pushed me toward it, and to this day, I regret not speaking up sooner. The decision had to be made quickly, but that doesn't excuse the impact. Under pressure, my behavior didn't align with my values, and the team noticed immediately.

They told me, "This goes against everything you've worked with us on." They were right. In that moment, I had a choice: defend the decision and hide behind my title, or own it. I chose to own it. I told them, "You're right. I apologize. This was something I said I'd never do, and I did it. That's on me." Then I said, "We all know the problem we're trying to solve. If this is the wrong solution, let's reset and find the right one together." We spent about an hour talking it through, exploring options, drawing on their context, and finding a better path. It fit the team. It fit the organization. And it was something we could all stand behind.

Afterward, I went to the C-suite and told them that the original solution I'd proposed was wrong. I explained that the team and I had worked together to find a better approach, and I made it clear we'd get a better outcome as a result. That's courage: admitting you were wrong, changing course, and giving credit to the team, not just yourself. That moment mattered because the apology wasn't the end of leadership. It was the reset. Owning the mistake reopened the conversation, restored trust, and made better thinking possible.[4]

---

4 Research on humble leadership suggests that leaders who admit limits, appreciate others' strengths, and stay open to learning help create stronger team climates and better performance. See Bradley P. Owens and David R. Hekman, "How Does Leader Humility Influence Team Performance? Exploring the Mechanisms of Contagion and Collective Promotion Focus," *Academy of Management Journal* 59, no. 3 (2016): 1088–1111.

And here's the part I don't want to skip: in that moment, the team showed more leadership than I did. They challenged me. They held me to the standard I'd set. That's exactly what you want. The strongest teams don't just follow leaders. They hold them accountable, too.

Humility

Humility is accepting that you don't have all the answers and you're not supposed to. This isn't the 1950s, where the supervisor is the oracle. There's too much complexity and too much information. No one person can hold it all. If you cling to the idea that "because I'm the leader, I should know everything," you'll make worse decisions, shut down your team, and destroy trust. Humility means remembering that your title means little without the team behind you. It means being willing to be wrong in public. It means being open to being challenged by people "below" you on the org chart. It means treating other people's insight as an asset, not a threat.[5]

Learning from Failure (Without Blame)

Failure will happen. Authentic leadership isn't about avoiding every mistake, pretending everything is fine, or finding someone to blame when it isn't. It's owning your part, looking for root causes rather than scapegoats, and asking "What can we do differently next time?" rather than "Whose fault is this?"

That shift matters more than most people realize. The moment a team believes failure will be used as a weapon, people stop telling the truth. They hide risks, soften bad news, protect themselves, and wait too long to speak up. That doesn't create accountability. It creates silence. When leaders take responsibility, share what they've learned, and invite the team into that learning, they build trust, resilience, and a culture

5 Ronald A. Heifetz distinguishes between authority and leadership, arguing that leadership is less about having all the answers and more about mobilizing people to face difficult challenges together. See Ronald A. Heifetz, *Leadership Without Easy Answers* (Cambridge, MA: Harvard University Press, 1994).

where people aren't afraid to try. When they point fingers, hide their own mistakes, and punish people for telling the truth, they build fear, silence, and a culture where everyone plays it safe, and nothing really changes.[3]

Authentic leadership isn't about being the hero. It's about building other leaders, admitting when you're wrong, and turning failure into fuel for learning, not weapons for blame. Those are the hard lessons. Most of us don't learn them from a book or a course. We learn them the hard way, by screwing up, listening, owning it, and choosing to do better next time.

Pulling It Together: How to Lead (Without the Hero Cape)

We've explored what leadership really is, what it looks like in chaos, and why courage and humility matter more than pretending you have all the answers. Now it's time to get practical and honest.

Manager Versus Leader: Which One Are You Being?

Your business card might say "manager," but every day you choose how you operate. That choice shapes everything. A manager-first approach sounds like this: my job is to have the answers. People bring me problems; I give them solutions. My authority comes from my title. When chaos hits, I need to control more and pull decisions up to me. If something goes wrong, I need to find out who screwed up. I measure success by compliance and hitting targets, even if it means pushing people hard.

A leader-first approach flips that script: my job is to create the conditions for others to find answers and grow. People bring me problems; I ask questions, listen, and help unblock them. My authority comes from trust, behavior, and credibility, not just my title. When chaos hits, I support the people closest to the work. If something goes wrong, I own my part, and we learn together. I measure success by how the team learns and adapts, not only by whether we hit the number. The difference isn't just semantics. It shapes how you show up, how your team responds, and how resilient your organization becomes.

Most of us move back and forth between these modes. This isn't about labeling yourself a "good leader" or a "bad manager." It's about noticing your default under pressure. Do you become more controlling, more defensive, more certain than the situation deserves? Or do you create the conditions for other people to think clearly? That's the real test. Not what you say about leadership when things are easy, but how you behave when the pressure is on, the stakes feel high, and your instincts start shouting for control.

There's a modern version of the same idea, too: teams take their cues from leaders. When you're regulated under pressure, you help the group make sense of what's happening. When you're reactive under pressure, you spread chaos. Leadership presence isn't about control or quick fixes; it's about staying grounded enough to tell the truth, create clarity, and help others do the same.[6]

Everyday Leadership (Without a Fancy Title)

Leadership is a practice, a set of behaviors anyone can choose every day. Sometimes, everyday leadership is as small as being the person who says, "I think we're solving the wrong problem," before the team burns a week on the wrong work. Sometimes it's noticing a teammate is overloaded and stepping in before things break. It doesn't have to be dramatic to matter.

Everyday leadership looks like raising risks early, even when it's uncomfortable. Helping teammates without waiting to be asked. Asking, "What do you need from me?" when tension rises. Pushing back respectfully when a decision doesn't make sense or feels rushed. Sharing information openly so people don't have to guess. Owning your mistakes and making it safer for others to own theirs, too. When more

6 This echoes the leadership insights of Xenophon, an ancient commander who emphasized stabilizing the emotional state of those he led during times of uncertainty and stress. Modern neuroscience supports this, showing that effective leadership under pressure involves regulating one's own emotional state to create psychological safety, foster trust, and enable the team's collective intelligence to emerge. This approach moves away from control and quick fixes toward presence, clarity, and emotional regulation as key leadership capabilities. Sources: Xenophon (*Anabasis*), Daniel Kahneman (*Thinking, Fast and Slow*), Robert Sapolsky (*Behave*), Antonio Damasio (*Descartes' Error*), and Karl Friston (predictive processing papers).

people lead this way, you don't just get better results. You build a culture where people feel safe, valued, and trusted to bring their best. That's behavior-first leadership in action, grounded in courage, humility, and trust.

Self-Reflection: When the Next Time Chaos Hits

To make this stick, here are a few questions to ask yourself after the next round of turmoil, or when you look back on the last one. When chaos hit…

- What happened to you?
- Did you get calmer, or more frantic?
- Did you create clarity, or did you create noise with meetings, messages, and thrashing?
- Did you ask the team what they thought, or did you jump straight to giving orders?
- Did you trust the people closest to the work, or did you pull everything up to your level?

Then zoom in on what was actually in your control. What was outside your control, no matter how hard you pushed? And what could you do differently next time that's actually within your influence? After the chaos was over, what did you do with it? Did you own your part, or did you quietly blame others? Did you make time to learn from it, or did you move on and hope it doesn't happen again? And what's one behavior you're willing to change before the next round hits?

You don't become a better leader by memorizing leadership quotes. You become a better leader by noticing how you behave when things get hard and then having the courage and humility to do something different next time. Leadership is where all of this either comes to life or dies on the spot. Some of the best leadership lessons don't come from leadership books at all. They show up in places where feedback is immediate, ego gets exposed fast, and adaptation is nonnegotiable.

For me, that's Jiu-Jitsu and puzzles. Both are live-fire labs for agility. You never have perfect information. Your plan rarely survives first contact. And the only people who get better are the ones who stay calm, experiment, and adjust. Think about what happens when someone new rolls (spars) in Jiu-Jitsu. The moment pressure shows up, they tense up, force bad moves, burn energy, and usually make things worse. They're trying to win the moment instead of understanding it. More experienced people tend to do the opposite. They breathe. They pay attention. They adjust. They stop fighting reality long enough to work with it.

Leadership under pressure isn't that different. The less secure you are, the more tempted you are to force control, talk more, tighten your grip, and react before you've really understood the problem. The more grounded you are, the more likely you are to notice what's actually happening, respond with purpose, and create enough calm for others to think, too. That doesn't make you passive. It makes you effective. Whether you're leading a team, solving a delivery problem, or trying not to get folded into a pretzel, the pattern is surprisingly similar: stay calm, pay attention, adapt, and don't let your ego make the situation harder than it already is.

In Chapter 8, we're going to step outside the usual business context and look at Jiu-Jitsu, puzzles, and the art of adaptation: how those worlds mirror real agility, why adaptability and creative problem-solving beat rigid plans, and what they can teach us about leading and working in complex environments without losing our minds.

CHAPTER 8

JIU-JITSU, PUZZLES, AND THE ART OF ADAPTATION

We've spent a lot of time talking about business, frameworks, and teams. Now I want to step off the whiteboard and onto the mat. Since 2008, I've trained and competed in Jiu-Jitsu with the Gracie Humaitá team, and I've always loved puzzles. That's probably why I'm tying them together here. Both normal puzzles with pieces and mental puzzles require a lot of thought. This includes the puzzle of a match where I'm trying to avoid getting tapped out or having my clothes folded by my opponent while I'm still wearing them.

For me, agility lives in how we think, move, and adapt. And Jiu-Jitsu is one of the clearest places I've seen real agility in action. Boardrooms love certainty, plans, and control. But you can change all the processes you want at the team level; if you don't change how the boardroom behaves, the organization doesn't really change.

On the mats, none of that matters. In Jiu-Jitsu, we say, "The mats don't lie." You can talk a big game. You can rehearse your favorite moves in your head. But the moment you roll with someone, reality shows up. Either your timing works, or it doesn't. Either you adapt, or you get swept, passed, or choked. Jiu-Jitsu and puzzles don't care about your plan or your title. They only care about how you respond when reality doesn't match your expectations. Jiu-Jitsu taught me to stay calm when the plan falls apart. Puzzles taught me something similar, just in a different arena: when you don't have the whole picture, you don't win by forcing harder. You win by seeing patterns, testing moves, and adapting fast.

So let's stay on the mats for a moment, because Jiu-Jitsu is a surprisingly clear model for how real problem-solving works in messy, volatile, uncertain, complex, and ambiguous environments.

No Plan Survives First Contact

If you've done Jiu-Jitsu, you'll recognize this immediately. If you haven't, here's how it usually goes. Before a roll or a competition match, I'll have a plan in my head: pull guard, hit a sweep, then set up a submission. Then the ref says "go," and that plan starts dying almost instantly because the other person has a plan too. As Mike Tyson once famously said, "Everyone has a plan till they get punched in the mouth." Your opponent doesn't move the way you imagined, and they don't give you the positions you expected. The mats don't care what you intended. They show you what actually works.

It's exactly like a product roadmap, a transformation plan, or a beautiful strategy slide deck. They're useful right up until reality hits. Agility isn't about having no plan. It's about the loop: plan, make contact with reality, adapt fast (see Figure 8.1). If you cling to the original plan after reality has clearly changed, you're not being "disciplined." You're just getting choked out more slowly.

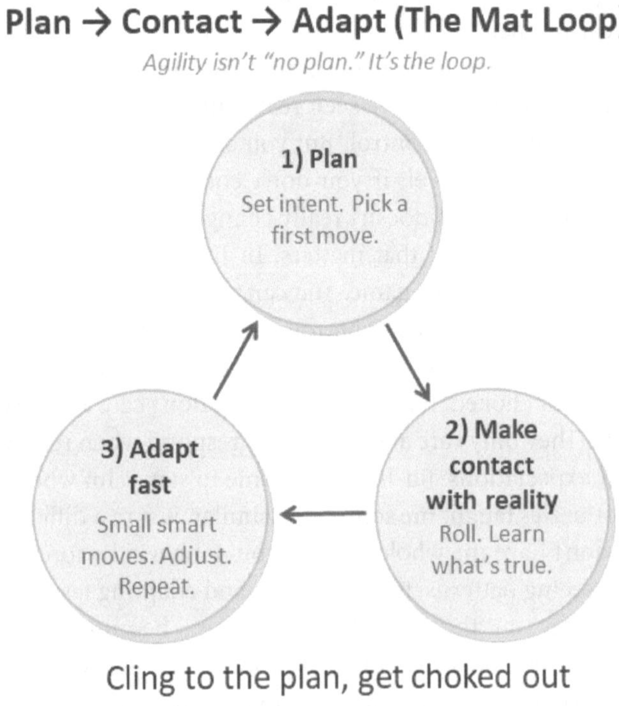

Figure 8.1 | Plan, contact, adapt

Pressure Versus Panic

One of the first things you learn in Jiu-Jitsu is how to handle pressure. You get stuck in tight spots, someone's weight is on your chest, your breathing gets shallow, and your body starts sending the same message over and over: panic. This is one of the clearest places Jiu-Jitsu has taught me what real agility looks like. Under pressure, you have a choice. You can thrash, burn energy, and try to force your way out, or you can breathe, stay calm, and look for small, smart movements. On the mat, a hip shift, a better angle, or a grip change can completely change the situation. You conserve energy, wait for the right moment, and escape or reverse when it's there. Freak out, and you might escape once, but you won't last the round. Teams under pressure face the same choice.

Panic looks like dropping everything for the latest "urgent" request, starting five things at once just in case, or piling on overtime and heroics until people burn out. Calm under pressure looks like getting a clear view of what's actually happening, asking what the smallest move that helps right now would be, and deciding what can be stopped so there's room to respond (see Figure 8.2).

Pressure vs. Panic (Mat ↔ Work)

	On the Mat	At Work
Signal	Tight chest, shallow breathing Tunnel vision Energy spikes	"Everything is urgent" Constant pings and meetings People go quiet
Panic looks like	Thrash and force Burn energy fast Make it worse	Start five things at once Overtime and heroics More chaos, less clarity
Calm looks like	Breathe and settle Small adjustments Wait for openings	Clarify what's true Stop something to make room One smart move at a time
Better looks like	Improve position first Use leverage, angles, timing	Reduce WIP, tighten feedback Make trade-offs explicit

Pressure is manageable. Panic is expensive

Figure 8.2 | Pressure versus panic

You've seen the difference before. Pressure is challenging but manageable. You're stretched, but you can still think, and when you succeed, you feel proud and energized. Panic is when the load never lets up, the goal feels impossible or unclear, and even when you "win," you mostly feel relief, followed by dread about the next round.

Jiu-Jitsu trains you to live under pressure without tipping into panic. You learn to recognize the signs: tight chest, shallow breathing, tunnel vision, and respond with control instead of flailing. Agile teams need the same skills.

Agility isn't about removing pressure. It's about handling pressure without turning it into chronic stress and panic. It's leaders and teams saying, "Yes, this is intense, but we're going to slow down, get clear on reality, make one smart move, and then another," instead of, "Everyone run faster and hope for the best."

Tapping: Failure as Feedback

In Jiu-Jitsu, you tap when you're caught in a submission, and you know you'll get hurt if you don't. In a good gym, tapping isn't shameful. It's feedback. It means you hit a limit, learned something, and get to keep training. In many workplaces, we do the opposite. We hide mistakes, pretend we're fine, and avoid admitting when we're stuck. We avoid tapping at all costs.

Healthy agility means building a culture where people can tap out, say, "I was wrong," "This isn't working," or "I need help," without getting punished. A healthy team treats tapping early as a sign of professionalism, not weakness. When you get tapped or stuck in a bad position on the mat, it's tempting to label it as failure. But as my coach likes to remind us, "You didn't fail. You hit your limit. Now build from that and use it."

I've lived this the hard way at work, too. I once accidentally wiped out about two-thirds of a test database with a single keystroke, pure accident, and it's why I still twitch when someone mentions COBOL. I could've tried to hide it. Instead, within minutes, I went to my manager, explained what happened, and offered a recovery plan. Because we had backups, we restored about 80% of the data quickly and resolved the rest with additional effort. If I'd tried to hide it, it would've taken longer to discover, been harder to fix, and trust would've taken a much bigger hit. Admitting the mistake early made it recoverable.

Different Partners, Different Games

In Jiu-Jitsu, you train with different sizes, strengths, styles, belt levels, and genders. You learn fast that the move that works on a brand-new white belt won't work on a seasoned black belt, and what works on a strong, pressure-heavy opponent might fail on someone smaller, faster, and more technical. You can't have one game for everyone.

It's the same in business. Every team, every product, and every organization is a different opponent. If you only have one move, one framework, or one way of working, you're in trouble. Personally, I'm on the heavier side of the weight classes. I can't roll the same way with someone my size or bigger as I do with someone much lighter and faster. If I try to use the same game for everyone, I end up losing a lot. I need different strategies for different bodies, different approaches for different belt levels, and different timing depending on what's in front of me. That's precisely why frameworks are tools, not religions. A good toolbox gives you options based on the situation. In Jiu-Jitsu, that's how you grow. In agility, it's the same. Frameworks, like techniques, are there to serve adaptation, not replace it.

Puzzles: A Model for Problem-Solving in a VUCA World

We're not working in a calm, predictable world anymore. Call it VUCA (volatile, uncertain, complex, and ambiguous), but it points to the same reality: a lot is going on, variables keep changing, and you can't keep everything neat, linear, and controlled.[1]

You are not in control of everything, and we need to accept that. That's where puzzles come in. Unlike a straightforward problem with one clear solution, puzzles force you to work with partial information. You have to see patterns, test moves, and adapt as new pieces come into view. You don't win by forcing a solution. You win by experimenting, learning from each attempt, and adjusting your approach.

1 The term VUCA originated in U.S. military education in the late 1980s and 1990s to describe the changing post–Cold War environment and has since been widely adopted in business and leadership literature. See, for example, Bob Johansen, *Leaders Make the Future: Ten New Leadership Skills for an Uncertain World* (San Francisco: Berrett-Koehler, 2009).

Constraints don't just limit you; they force adaptation. A puzzle's rules and pieces define the playing field, but within those limits, creativity and strategy thrive. Just like on the mats, where you can't control your opponent, puzzles teach the same lesson: success comes from understanding the environment, recognizing what's possible, and making smart moves step by step, even when the whole picture isn't clear.

In agility, that ethos matters. Frameworks, like puzzle-solving techniques, are tools to help you navigate complexity, not rigid scripts to follow blindly. That's one of the reasons I love puzzles. Outside of Jiu-Jitsu, they're my model for problem-solving. With any puzzle, jigsaw, logic, whatever, you never start with the complete solution. You start with edges, patterns, colors, and small clusters.[2] It's the same as real work: we never know the whole picture up front. We only know what we know right now, so we start with what's visible and build from there. We look for small wins that lead to something bigger later. You find a pattern. You see how pieces fit. You turn things around until something clicks.

So why do we suddenly forget this at work? Why do we force a framework, force a deadline, or force a solution that clearly doesn't fit just because it's on a slide deck? Why do we behave differently at work than we do in almost every other problem-solving situation in our lives? It doesn't make sense. Real agility is being willing to say, "This doesn't fit. We misjudged this. Let's try a different angle."

Pattern Recognition, Patience, and Partial Progress

Puzzles ask three things of you: pattern recognition, patience, and your ability to hold partial progress in your head. When you're working on a jigsaw, you might finish a corner, then a cluster in the middle, then a strip of sky or trees. You don't go piece by piece in a straight line. You build what fits, when it fits. In a logic puzzle, you do the same thing in a different form. You rule out options, follow a thread, hit a dead end, and then reframe the problem with what you've learned.

2 Research on problem-solving and creativity suggests that effective solutions often emerge through iterative exploration, testing patterns, reframing the problem, and combining partial insights, rather than from a single, fully formed plan. See Keith J. Holyoak and Robert G. Morrison, eds., *The Cambridge Handbook of Thinking and Reasoning* (Cambridge: Cambridge University Press, 2005), especially chapters on insight and analogical problem-solving.

Agile work is the same. It's messy, nonlinear, and full of half-finished pieces. Sometimes you're working on something three steps ahead before the basics are fully polished, and that can be fine if it's intentional, visible, and the team is aligned on why you're doing it. You need people who can spot patterns in the chaos, get comfortable with the incomplete, and trust that the small wins will connect later. That's a very different ethos from "we need the full plan up front" or "everything must happen in a straight line." Real work is the same: you move forward with partial information and connect the pieces over time.

Collaborative Puzzles = Cross-Functional Teams

Puzzles also show you what real collaboration looks like. When multiple people work on a puzzle, different strengths show up right away. One person is great at finding edges. Another sees color patterns and textures. Another locks in on specific areas like faces, text, or buildings. People move around the table, share what they see, help each other when they're stuck, and don't compete for credit over who placed the most pieces.

The puzzle gets solved faster because strengths are used, information is shared, and help shows up where it's needed. It's not about silos, "that's not my job," or "I only do X." It's everyone contributing where they can make the biggest difference.

Think about a sports team. Each player has a position, but they move, support, and cover for each other. They don't stay in one lane and ignore what's happening everywhere else. In a healthy team, no matter the industry, people move around the puzzle. They help where it makes sense, share information instead of hoarding it, and care more about solving the problem than about who gets credit.

And this is where many so-called Agile environments fail. The focus shifts to the individual within the team rather than the team itself. We reward individual heroics, track individual output, and run performance reviews that focus on solo stats. Then we act surprised when people optimize for themselves, collaboration gets weird, and the puzzle takes longer and hurts more than it should.

If you want real agility, stop forcing bad fits and start treating work like a puzzle that only comes together through teamwork. Build teams that can see patterns, live with partial progress, and move around the table together. Better yet, review and reward teams as teams. Recognize group wins, shared learning, and the quiet support that makes everyone better. When you shift the spotlight from individual to team, collaboration becomes the norm, not the exception. That's where real adaptation happens.

Adaptability and Creative Problem-Solving Beat Rigid Plans

What does all this mean for agility? We need to double down on adaptability and creative problem-solving because they consistently outperform rigid plans. Whether it's Jiu-Jitsu or puzzles, the common thread is the same: you'll be wrong a lot, you'll be surprised a lot, and you can't brute-force your way to a good outcome. You can try to jam the wrong puzzle piece into place or muscle your way into a position on the mat. Technically, you might force something, but the puzzle will look wrong, the position will be unstable, and you'll burn way more energy than you should. One small shift, and it all falls apart.

It's the same in teams. You can force a framework, a deadline, or a solution, but you'll pay for it in stress, rework, burnout, and lost trust. The people and teams who do well are the ones who stay calm, stay curious, and adjust without turning every setback into drama. We're not that far off. The foundations for real agility are already there. What gets in the way are the small, stubborn things: ego, attachment to the original plan, fear of looking wrong, and obsession with control.

And let's be clear, this isn't a free-for-all. Every domain has constraints, governance, and regulations, and that's fine. Adaptability isn't about breaking the rules. It's about finding smarter ways to work within them.[3]

3 Research on creativity and problem-solving suggests that constraints can actually enhance innovation by focusing attention and forcing people to explore alternative paths, rather than relying on default habits. See Patricia D. Stokes, *Creativity from Constraints: The Psychology of Breakthrough* (New York: Springer, 2006), which argues that well-designed limits can drive more original solutions. For a broader view on working effectively within complex, constrained systems, see Dave Snowden and Mary E. Boone, "A Leader's Framework for Decision Making," *Harvard Business Review* 85, no. 11 (2007): 68–76, which introduces the Cynefin framework for navigating different types of contexts without pretending everything is fully controllable.

In Jiu-Jitsu, if you're stuck under someone heavier, you have limited options, but you still have options. You use leverage, timing, and angles to escape. In banking or other heavily regulated industries, you have strict rules and controls. You don't break them. You find ways to be more effective within them. In puzzles, if you're missing a piece, you don't quit. You work around it, build other sections, and when the missing piece shows up, you drop it in and keep going.

Business is no different. Limited budget? You prioritize and get creative. Legacy systems? You build around them, integrate smartly, and chip away over time. Regulations? You design processes that respect them and still let people move. Adaptation isn't about ignoring reality. It's about working with reality instead of pretending it's something else.

Ego Versus Adaptation

In Jiu-Jitsu, ego says, "I shouldn't have to tap to this person." Adaptation says, "Tap, learn, roll again." With puzzles, ego says, "I should be able to solve this fast." Adaptation says, "Slow down. Change your strategy. Look from another angle."

In leadership and teams, ego clings to the original plan, framework, or idea: "We said we'd do it this way, so we're doing it this way." Adaptation says, "Okay, that didn't work. What now? What are we learning? What needs to change?"

Ego is rigid. Adaptation is Agile.

Five Lessons from the Mat and the Puzzle Table

Let's tie this chapter together with five lessons that show up everywhere, from the mat to the meeting room.

1. *Stay calm under pressure.* Slow down enough to think before you make things worse.
2. *Treat failure as feedback, not identity.* Mistakes are information if people can tell the truth.
3. *Use the whole team's perspective.* Different people see different risks, patterns, and options.

4. *Don't force bad fits.* If the plan, tool, or framework clearly doesn't fit, stop pretending it does.
5. *Adaptation is a skill you train.* You get better by practicing under real constraints, not by talking about it.

You don't become adaptable by talking about agility. You become adaptable by practicing it. That means showing up under pressure, working with real constraints, and navigating the messy reality of teams and organizations. The more you build that muscle, the less you need to hide behind frameworks, buzzwords, or rigid plans to feel safe.

Everything we've discussed about behavior-first agility, team well-being, and leadership in chaos comes to life on the mat and at the puzzle table. Jiu-Jitsu and puzzles strip away the slides, titles, and jargon. There's no transformation program to hide behind and no framework to blame. It's just you: your habits, your ego, your ability to stay calm, spot patterns, ask for help, and try again.

That's the same muscle you need in real organizations: the willingness to adapt in the moment, work with constraints instead of complaining about them, and treat failure as information, not as a personal indictment. These metaphors aren't cute extras. They're practice grounds for the exact behaviors real agility demands.

So far, we've explored principles, stories, and metaphors, from graveyards of failed frameworks to Jiu-Jitsu mats and puzzle tables. Now we get into the evidence.

In Chapter 9, we'll dig into honest case studies: what worked, what blew up, and why. We'll celebrate the wins without pretending they were magic, and we'll own the failures without sugarcoating them. Most importantly, we'll learn from both, without ego, so you can recognize these patterns in your own world and make better calls when it's your turn in the arena.

CHAPTER 9

REAL-WORLD WINS (AND FAILS)

Let's walk through real case studies where things worked, where they absolutely didn't, and why. You might read these and draw your own conclusions, and that's the point. I'm sharing what it felt like from inside these situations, and what they taught me about what creates an Agile ecosystem and what kills it. These aren't sanitized success stories. They're messy, human examples where behavior, fear, ego, and courage all show up in real organizations.

Case Study 1: The Software Company That "Paused Agile"

You've already seen this pattern earlier in the book, so I won't retell every detail. A software company was missing delivery dates, struggling with quality, and falling behind client demand. Leadership said they wanted an Agile environment (see Figure 9.1). We assessed the end-to-end flow, spoke with teams and leaders, and recommended a behavior-first approach: leadership workshops, executive coaching, and team support to change how decisions were made, not just how work was tracked.

What we found wasn't unusual. Teams were trying hard, but they were working inside a system full of mixed signals. Priorities shifted without warning. Decisions got delayed or overridden. People were expected to move faster without being given more clarity, support, or authority. The teams didn't need more pressure. They needed a better environment.

Case Study Snapshot (At-a-Glance)

	Case 1: "Paused Agile"	Case 2: Went Agile
Starting situation:	Missed dates, quality issues, demand pressure.	Growth, shifting priorities, retention risk.
What they wanted:	"An Agile environment" without disruption.	Focus, execution, and better retention.
What they actually did:	Part-time team training; no leadership change.	Visible work, weekly decisions, student feedback loops.
Leadership behavior:	Avoided loss of control; delayed follow-through.	Owners adapted; used feedback over ego.
What happened:	Momentum died; system snapped back.	Hit 200 students early; kept growing sustainably.
Why (in one line):	They wanted teams to change, not themselves.	They treated agility as learning, not a label.

Figure 9.1 | Case study snapshot

Leadership chose the lowest disruption option: train a couple of teams for one day a week and keep everything else the same. No executive workshops. No leader coaching. No follow-through plan. The teams engaged and tried. Leadership paused, delayed, and eventually shelved the whole thing. What made it harder was that the teams weren't resisting. They were trying. They were showing up, experimenting, asking good questions, and doing their best to apply what they were learning inside a system that kept sending them mixed messages. One day, they were told to collaborate and self-organize more. The next day, decisions were still getting pulled back up the chain, priorities were still changing without warning, and the same old approval habits were still in play. That kind of environment wears people down fast. It teaches them that the language may have changed, but the power structure hasn't. And once people feel that gap between what leadership says and how leadership behaves, cynicism shows up quickly.

The result wasn't surprising: pockets of new behavior showed up briefly, then the system pulled everything back to the old default. What went wrong? The simplest answer is this: leadership wanted the teams to change, but not themselves. The deeper answer is that they wanted the benefits, better delivery, better quality, faster response times, without giving up control. They treated agility as a project they could pause rather than as a shift in behavior and ethos.

That distinction matters more than most people realize. Projects can be delayed, rescheduled, deprioritized, or quietly killed when something feels uncomfortable. Behavior change doesn't work like that. Once you start asking people to work differently, speak more honestly, surface problems earlier, and make decisions closer to the work, you're already changing the social contract. If leadership treats that like a temporary initiative instead of a real shift, people notice. They stop trusting the message. And when trust drops, even good practices start to look performative.

In the room, you could feel it. Fear of losing control. Fear of letting teams make decisions. Fear of changing how leaders operated. So they defaulted to the safest move: train the teams and see how it goes. That's a common pattern in organizations that say they want change but only want it at the edges. They want teams to collaborate more, adapt faster, and solve problems earlier, while leadership keeps the same approval structures, the same decision bottlenecks, and the same habits that created the mess in the first place. This is what Ronald Heifetz, a thought leader on adaptive leadership, points to when he argues that people don't fear change as much as they fear loss. In this case, it was the loss of control, status, and comfort with old habits.

If leadership isn't willing to change with the teams, you don't have an Agile transformation. You have framework performance. And over time, everything returns to normal, usually with more cynicism than before.

Case Study 2: The Jiu-Jitsu Club That Went Agile, for Real

Now for something you don't see every day: an Agile shift in a Jiu-Jitsu club. This was a growing club that had just moved into a larger location. There was a lot to do, not a lot of time, and priorities were changing constantly. They struggled with focus, frequent pivots, and turning good ideas into execution. They brought me in to look at how they worked, help them get focused, build a roadmap that could adapt, and improve retention, not just chase new signups.

At its core, agility is about feedback, learning, and understanding what customers actually need. People joined this club for different reasons: self-defense, fitness, learning a new skill, or competition training. The owners needed to know what was working, what wasn't, why people stayed, and why they left.

Their membership model was month-to-month, which is rare in martial arts, where many clubs lock people into long-term contracts or the infamous black belt contract.[1] This club wanted students to stay because they were happy, not because they were trapped. That made retention even more important, and it meant feedback loops had to be tight. So we kept it simple. We set up a basic Kanban system to make work visible. We ran weekly check-ins to review goals, clarify roles, and decide what mattered this week. And we built short feedback loops with students so they could adjust classes, communication, and offers based on real experience, not just gut feel.

We didn't use templates, jargon, or some big complicated operating model. We made the work visible, talked about it regularly, and adjusted based on what we were learning. Instead of carrying everything around in their heads, they could see what was going on. Instead of reacting to whatever felt loudest that day, they could talk through trade-offs and make clearer calls. Instead of guessing what students needed, they had more chances to hear it directly and respond while it still mattered. That kind of visibility lowers the noise. It gives people room to think. And when people have room to think, they usually make better decisions.

What made this work wasn't complexity. It was consistency. Work stopped living in people's heads. Priorities became visible. Conversations got shorter and clearer. Instead of reacting to everything at once, they

1 A "black belt contract" is a long-term membership agreement where you pay your fees upfront for what's supposed to be the time it takes to earn a black belt (often framed like paying for a five-year membership in advance). The pitch is that you can take as long as you want to earn the belt—but the money is committed either way. These kinds of long-term club-style contracts can also lead to consumer disputes when cancellation terms are unclear or when billing/collections are handled by third parties via contract assignment. See "Assignment Clauses in Health Club, Gym and Martial Arts Membership Contracts," www.thelaw.com/law/assignment-clauses-in-health-club-gym-and-martial-arts-membership-contracts.564/.

could see what needed attention now, what could wait, and what wasn't worth doing at all.

They set a clear goal: reach 200 students by the end of the year. They hit it in August. The gym kept growing. Students were satisfied. And their attrition rate was lower than most other clubs in the area. Why did this work? Because the owners were willing to change their approach. They used feedback instead of ego. They set a clear goal and used a simple system to move toward it. Most importantly, they treated agility as a means of learning and adapting, not a label to wear.

They also didn't overcomplicate it. That helped more than people think. A lot of organizations bury themselves in process because process feels safer than honesty. This club didn't need a giant operating model. They needed a way to see what was happening, decide what mattered, and adjust before small problems became bigger ones. That's one of the reasons the changes stuck. The system fit the reality of the business instead of forcing the business to fit someone else's template. No massive rollout. No 200-slide transformation deck. Just clear goals, visible work, short feedback loops, and a willingness to adjust. That's agility!

And there was something else underneath it: trust. They were willing to look honestly at what was and wasn't working. They didn't feel the need to defend every past decision. They didn't need to pretend every idea was good. They were willing to see reality, respond to it, and keep going. That's what made the system sustainable.

The Pattern Across Both Stories

And here's the meta-lesson across both stories: you have to be able to celebrate the successes and own the failures, not just other people's failures, but your own. On the surface, these stories look like opposites. One organization had more resources, more formal structure, and more obvious reasons to invest in change. The other was a martial arts club trying to grow without losing its grip on reality. And yet the smaller, simpler environment adapted better. Why? Because agility doesn't care how impressive your org chart is. It cares whether people can see reality, tell the truth, make decisions, and adjust their behavior when the facts change. That's the pattern that keeps repeating. Not industry. Not size. Not how many Agile words people know. Behavior (see Figure 9.2).

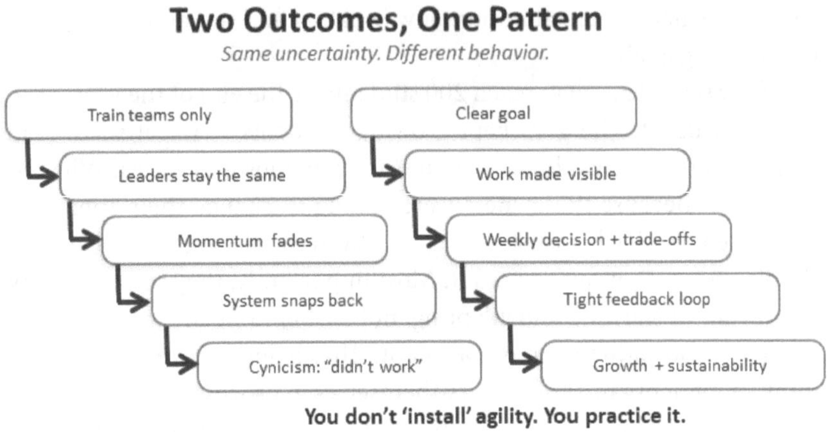

Figure 9.2 | Two outcomes, one pattern

Where I Failed the Software Company

In that first story, it's easy to point at leadership and say, "They didn't buy in," "They were afraid of losing control," or "They paused Agile and killed the momentum." All of that is true, but it's not the whole story. They weren't cartoon villains. They didn't fully understand what was at stake. They wanted to protect what they had. They had budget concerns and comfort zones. I can't say exactly why they made those choices, but I can take responsibility for what I was responsible for.

I delivered the team training and did what little coaching I could, but I failed those teams in a few key ways. I didn't push hard enough to help leadership understand the impact of ignoring most of the recommendations. I didn't make the risks clear: training only two teams at a time, stretching it over months, and not touching leadership behavior at all. And when they paused, then paused again, then killed the whole thing, I didn't do a good enough job making the value visible or making the risk of stopping clear.

From their point of view, it probably looked like this: We paid for training. Things didn't really change. So this Agile thing isn't worth it. That's on me. If I were walking into that same situation now, I'd handle it differently. I'd spend more time up front making the consequences visible in language that leadership couldn't ignore. I'd tie behavior change more directly to delivery risk, quality issues, employee frustration, and the cost of delay. I'd be clearer that partial adoption wasn't a cautious version of transformation; it was a high-probability path to disappointment. And I'd push harder to make leadership ownership nonnegotiable because if the people with the most power to shape the environment stay untouched, the rest of the effort is mostly theatre with better vocabulary. Figure 9.3 provides a great way to review your own behavior.

This is a common failure for coaches and change leaders: it's hard to show value when your work is mostly intangible. You're not shipping a product. You're helping people think and work differently, and that's harder to see, especially for leaders who are used to dashboards, deliverables, and clean ROI. At the time, I was still assuming that if people experienced enough of the benefit, they'd connect the dots. Sometimes they do. Sometimes they absolutely do not. If the value isn't visible, many leaders will default to what they can count, what they can control, and what feels safer in the moment.

Self-Accountability Checklist

Before you blame "resistance," check your own work.

- ☐ I made the risk of not changing visible.
- ☐ I made the cost of stopping visible.
- ☐ I tied behavior change to real outcomes.
- ☐ I pushed for leadership involvement early.
- ☐ I showed value in a way leaders understand.
- ☐ I named trade-offs instead of selling certainty.
- ☐ I didn't assume "they'll just get it."
- ☐ I owned my part when it didn't land.

If you can't show value and risk,
leaders will default to comfort.

Figure 9.3 | Self-accountability checklist

Over time, I had to learn how to fix that. I went back to old MBA materials, articles, and books on ROI, as well as research on measuring intellectual and behavioral value. I started asking different questions: How do I demonstrate return when the work lives mostly in people's habits? How do I make the risk of not changing visible? There are tools out there. Sometimes you adapt them. Sometimes you build your own. That failure forced me to get better at demonstrating value, making risk and ROI clearer, and stopping the habit of assuming they'll just get it. That's the biggest learning I took from that mess.

The Jiu-Jitsu Club Surprise

The Jiu-Jitsu club story is the opposite. The original goal was modest: help them get organized, clarify what they were doing, and make sure everyone knew the goals, who was doing what, and how it connected back to the students. That's what they wanted. That's what I expected. And then something bigger and better happened. They made work visible. They built real transparency. They tightened feedback loops. They hit 200 students months early and kept growing at an impressive pace for a young club. They added more classes, expanded into judo, self-defense, and personal training, built out a gym, and grew membership without it taking over their lives.

And that's the part that matters most. The real success wasn't just hitting the number. They built a way of working that supported their lives instead of consuming them. They could grow without burning out. They could adapt quickly because work was visible, decisions were shared, and feedback was constant. That's the point. A good Agile environment isn't just about getting more done. It's about making better use of time so people can have a life outside of work. That's the kind of success we should be aiming for.

The funny part is that this kind of success often gets overlooked because it doesn't look dramatic enough. There's no giant launch. No glossy transformation story. No executive panel talking about disruption. Just a team getting clearer, calmer, and better at responding to reality. But that's usually what real progress looks like. It looks ordinary from the outside and life-giving from the inside. Less chaos. Better decisions. More room to breathe. That's not boring. That's the win.

Ego Versus Adaptation: The Pattern That Keeps Repeating

Ego plays a huge role in all of this. If you let it drive you—thinking, I'm the best change leader in the room, or I know what's right, they just don't get it—you're going to run into problems. But the opposite ego trap is real, too. If you're constantly second-guessing yourself, convinced you're not good enough, and never speak up about value or risk, you'll run into problems as well.

Everyone has an ego. That's not the issue. The real question is whether you can put it aside long enough to admit when you screwed up, learn from it, and try a different approach next time. Perfection doesn't exist here. The game is simple: get 1% better per day. Keep moving. There isn't a single person in the Agile space who can honestly say they're perfect at this. If they do, flip back to Chapter 3, and you'll know exactly what bucket to put them in.

Real agility and authentic leadership mean celebrating the wins, owning the failures, and letting both change how you show up next time. These case studies aren't about good companies and bad companies. They're about behavior, fear, ego, and follow-through.

The software company showed what happens when leaders want the benefits of agility without changing themselves: training becomes a checkbox, momentum dies, and everything drifts back to normal. The Jiu-Jitsu club showed the opposite: clear goals, short feedback loops, visible work, and owners willing to adapt. Same world. Same uncertainty. Very different outcomes.

The real lesson isn't to do Kanban or run workshops. It's this: celebrate the wins, own your part in the failures, and use both as fuel to improve how you lead, coach, and change systems. The tools were never the point. The point was whether the people involved were willing to see reality, tell the truth about what wasn't working, and change their behavior accordingly. One group wanted the appearance of change without the personal cost of changing. The other was willing to adapt in small, practical ways and keep learning as they went. That's why one story collapsed and the other kept growing. Not because one had better jargon, but because one had better follow-through.

And that sets up the next big question: how do you keep getting better without turning your organization into a never-ending experiment that exhausts everyone? Too many Agile environments confuse continuous improvement with constant churn, new tools, new structures, and new slogans every quarter.

In Chapter 10, we'll dig into continuous improvement (not continuous chaos): the difference between real improvement and pointless change, how to build habits and systems that are sustainable, and practical ways to keep evolving without burning people out along the way.

CHAPTER 10

CONTINUOUS IMPROVEMENT (NOT CONTINUOUS CHAOS)

Continuous improvement is one of those things everyone says they want. Better products. Better workflows. Better leadership. Better lives outside of work. So why is it so hard to sustain? Why do we improve only to end up right back where we started? A big part of the problem is simple: we don't distinguish between real improvement and pointless change. Let's fix that.

We also get tripped up because we talk about improvement in terms of attributes instead of defining it. We say things like "more Agile," "more innovative," or "more aligned." But those are vibes, not definitions. A definition draws a boundary. It tells you what something is, and what it isn't. So here's mine: *Real improvement is a change that reduces friction or increases value without creating new pain somewhere else.* If people can't feel it in the flow of work, the quality of decisions, or the sustainability of the pace, it's probably just activity (see Figure 10.1).[1]

1 In philosophy, a *definition* draws a boundary around what something *is* (and isn't). An *attribute* describes qualities something might have. For example, saying "agility is adaptive and fast" lists attributes. Still, it doesn't define agility—because lots of things can be adaptive or fast without being "Agile" in any meaningful sense. This matters in organizations because when we rely solely on attributes ("more aligned," "more innovative," "more Agile"), we can't test whether a change actually worked. Clear definitions make improvement measurable and falsifiable: you can check whether friction went down, flow improved, or outcomes got better—rather than debating vibes.

Real Improvement vs. Pointless Change (the Test)

Does it save time or reduce friction without creating new pain elsewhere?

	Real Improvement	Pointless Change
What it looks like	➢ Removes a bottleneck ➢ Fewer stops, fewer handoffs ➢ Work moves faster, calmer	➢ New labels, new templates ➢ Reorgs and reshuffles ➢ More motion, same pain
What people feel	➢ "Oh nice, this is easier" ➢ Less waiting, less rework ➢ More focus, less thrash	➢ "Why is this harder now?" ➢ More clicks, more approvals ➢ Confusion, cynicism
Signals/Measures	➢ Cycle time drops ➢ Fewer defects/escalations ➢ Burnout signals trend down	➢ Vanity metrics look better ➢ More dashboards, less action ➢ Same fires, new vocabulary
Examples	➢ Remove pointless approvals ➢ Automate a boring check ➢ Make printing one click	➢ "95% pass" dashboard alone ➢ Cosmetic feature creep ➢ Rebrand; new paint, same pain

Figure 10.1 | The distinction between real improvement and pointless change

Real Improvement: Improvement = Saved Time (and Reduced Friction)

If you strip it down, real improvement is simple: it saves time or reduces friction—for you, your team, your organization, or your customers. That's it. Not more meetings. Not new labels. Not "we reorganized the org chart again." Real improvement shows up as less wasted time and less unnecessary pain.

Think about a clunky internal process. Every request has to go through three approvals, two spreadsheets, and a mysterious email chain before anything happens. Work piles up. People wait. Everyone gets frustrated. When you remove that bottleneck and requests move in hours instead of days, you've improved something. You haven't just changed it—you've saved time and removed friction. The same people, doing the same kind of work, can now get more done with less hassle.[2]

2 Many continuous improvement approaches, from Kaizen in lean manufacturing to modern UX research, emphasize small, incremental changes that remove waste and friction rather than dramatic overhauls. See Masaaki Imai, *Kaizen: The Key to Japan's Competitive Success* (New York: McGraw Hill, 1986), which frames improvement as ongoing removal of "muda" (waste) in processes, and Don Norman, *The Design of Everyday Things* (New York: Basic Books, 2013), on reducing "gulf of execution" and "gulf of evaluation" in user interactions.

It's like clearing a clogged filter. The system is the same. The work is the same. But suddenly things move the way they should: less resistance, fewer weird slowdowns, and less "why is this so hard?" The same logic applies to products, services, and the experience of both internal and external clients. Maybe your product has a form builder. Customers fill out a form and then, almost every time, print it manually afterward. You see the pattern in analytics and support conversations. Instead of shrugging, you add a simple step: when they finish the form, you ask, "Do you want to print this now? Yes/No." If they select "Yes," it prints.

You haven't reinvented the product. But you've saved them a few steps every single time, reduced friction in a workflow they repeat constantly, and created a small "oh nice" moment that adds up over weeks and months. It's the same with services. Maybe your internal clients, another department, have to jump through hoops to get what they need from you. If you spot the pattern and remove a pointless approval or automate a manual step, you're making their day easier and freeing up time for everyone.

That's continuous improvement: small, targeted changes that save time, reduce friction, and make life easier for the people doing the work or using the service. Real improvement always comes back to a few questions: What are we making easier? Where are we saving time or reducing pain? What value are we adding for the people using this? If you can't answer those in concrete terms, you're probably not improving. You're just changing things and hoping it counts.

Pointless Change: Motion Without Meaning

Pointless change is change for its own sake. It looks busy. It feels like "we're doing something." But underneath, nothing important gets better. It's motion without meaning. Take dashboards. One of the most pointless metrics I see in software projects is a simple pass/fail rate for testing. On its own, it tells you almost nothing. You know tests are being run, some pass, some fail—and that's about it. You don't know where the real risks are, whether quality is improving, or whether you're catching issues earlier or just pushing them downstream.

If you combine that data with trends over time, severity, and which parts of the system are affected, you can start to tell a story. You can see whether failures are clustering in a risky module, or whether critical defects are dropping as you improve your pipeline. But "95% pass, 5% fail" on a dashboard by itself? That's not improvement. That's noise, a vanity number that creates the illusion of control without changing any behavior.[3]

Cosmetic feature creep is another flavor of pointless change. Every year, a new device or app version ships with a "what's new" list packed with emojis, sticker packs, dark mode, and custom skins. Fun? Sometimes. But do they save meaningful time, reduce real friction, or solve an actual problem for customers? Sometimes yes. Often no. A lot of this work starts as "let's make it better," then quietly drifts into "let's make it look busy."

Novelty isn't automatically bad. The problem starts when novelty becomes a substitute for improvement. It's easier to ship, easier to market, and easier to point to in release notes than it is to deal with the friction everyone actually feels. You see it in user interface (UI) redesigns, too. A company decides to "refresh" the interface, moves core buttons and menus, adds extra clicks "for clarity," and forces everyone to relearn basic actions. Users' existing workflows and muscle memory don't get a vote. The result is predictable: people are slower, more frustrated, and can't find what they used to do in seconds. You can write all the release notes you want, but most people won't read them. And even if they do, they still have to retrain their hands. If your improvement makes experienced users worse at their job for weeks or months, it's not an improvement. It's churn.

And it's not just a product problem. Services and internal processes fall into the same trap. A department "improves" its workflow by adding an approval step or changing a form, without noticing it creates a bottleneck or dumps extra work onto another team. What saves a

3 Vanity metrics and poorly designed KPIs can create a false sense of progress while driving little real improvement. See Eric Ries, *The Lean Startup* (New York: Crown Business, 2011), on the distinction between "vanity metrics" and "actionable metrics," and John Doerr, *Measure What Matters* (New York: Portfolio, 2018), for examples of how misaligned metrics can distort behavior instead of improving outcomes.

few minutes for one group adds hours of frustration for another. The net result is more friction, not less. Rebrands are another classic: new logo, new tagline, same broken experience. Support doesn't change. The product doesn't change. Decision-making doesn't change. All that changes is the paint. That's not transformation. That's relabeling.

And then there's policy whiplash. Leaders confuse "being responsive" with "changing direction constantly." Strategy flips three or four times a year. Remote work policies change every few months. Tools and processes get swapped so often that no one can settle into a rhythm. Teams stop planning because "it'll change again anyway." People stop believing that direction is real, and most of their energy goes into reacting to the latest policy rather than improving how work is done.

Continuous improvement is not constant churn. It's not ripping up the floorboards every quarter because someone is bored or wants to look decisive. Real improvement is iterating on what matters, steadily, in ways that actually save time and reduce friction for real people, whether they're customers, clients, or your own colleagues. If it doesn't save time or reduce pain, it's just churn with a meeting invite.

How to Build Sustainable Systems (Without Getting Stuck)

How do we move from pointless change and continuous chaos into real, continuous improvement? Not with a magic template. Not with "the one system" that works everywhere. We get there by building systems where improvement is expected, welcomed, and used—and where people can tell the difference between real improvement and random change.

You need continuous improvement baked into your culture, not just your process.[4] Tools, boards, and events can help, but they're not the point. If the underlying beliefs and behaviors don't change, the tools just become a new framework for performance. That means you don't push

4 Many lean and quality movements emphasize that continuous improvement is primarily a cultural and behavioral practice, not a set of tools. W. Edwards Deming argued that most performance issues are systemic and require changes in management philosophy, not just local fixes. See W. Edwards Deming, *Out of the Crisis* (Cambridge, MA: MIT Press, 1986), especially his discussion of "constancy of purpose" and building a culture of ongoing improvement.

change for the sake of change. You prioritize sustainability, not constant turnover. You build systems where improvement sometimes shows up in the product, sometimes in how you work, and sometimes in how you treat each other.

If you create psychological safety (Chapter 6), trust and real teamwork (Chapters 5–7), and clarity about what "better" actually means (this chapter), you'll start to see consistent improvement—even if it doesn't look flashy on a roadmap. That's the difference between continuous improvement and continuous chaos.

And this goes deeper than process tweaks. It's about the true nature of agility: your ethos, your values, your character, and what you're actually trying to make better for customers and for the people doing the work. Here's the important part: every organization and every team is different. It would be ridiculous for me to pretend I can give you one system that works for every context. That would be nonsense, and a disservice to you. There are lots of ways to implement continuous improvement effectively. I've written a whole book (*Root Cause Analysis: True QA*) that goes deep on how to create that kind of environment. I'm not going to cram a fake "one-size-fits-all" solution into a single chapter here.

What I can give you is what we've been building toward: a straightforward way to tell real improvement from pointless change, and a lens you can use to design systems that create positive impact for teams, make work better for individuals, improve how departments and organizations function, and actually help your clients and customers. That's how you build something sustainable: not by copying someone else's blueprint, but by using your collective intelligence to design your own. Here's the truth:

- You know your organization the best.
- Your teams know where the friction really is.
- Together, you already know a lot about how to improve; you just need the space and safety to act on it.

If you want one practical principle for building sustainable systems, it's this: Work with the people doing the work. That sounds obvious, but many organizations still don't do it. Those frontline workers know where the real problems are. They know which changes will actually help. They know how to stay within regulations and constraints while still improving their work lives. High-performing organizations treat that type of knowledge as a core asset, not an afterthought.[5]

This is where Lean got something right: go to the Gemba, go where the work actually happens.[6] Not to audit people. Not to "check compliance." But to learn and to see the friction, the handoffs, the waiting, the workarounds, and the quiet heroics that never show up in a status report. When you bring that collective intelligence together, you get better decisions than any one layer of the organization could make alone. That's where sustainable continuous improvement comes from: not from a framework, but from people who are trusted to make things better.

In the next section, I'll share some practical ways to keep improving without burning out. But remember: those are examples, not commandments. The real power is in how you and your teams adapt them to your world.

Practical Ways to Keep Improving Without Burning Out

How do we keep improving without exhausting everyone? Figure 10.2 shows a few practical ways to make life easier for you, your team, and everyone around you—without turning "continuous improvement" into "continuous exhaustion."

5 Research on high-reliability and high-performing organizations consistently shows that involving the people closest to the work in problem-solving and improvement leads to better outcomes and more sustainable change. See Steven J. Spear and H. Kent Bowen, "Decoding the DNA of the Toyota Production System," *Harvard Business Review* 77, no. 5 (1999): 96–106, which describes how frontline workers at Toyota are expected and empowered to identify problems and improve processes.

6 Gemba (現場) is a Japanese term often translated as "the real place." In Lean thinking, "go to the gemba" means going to where the work actually happens (the shop floor, the support queue, the clinic, the codebase, the dojo mat—whatever "real" is in your context) to observe, learn, and understand the system before trying to improve it. It's not about surveillance or blaming people—it's about seeing reality up-close so you fix the right problems.

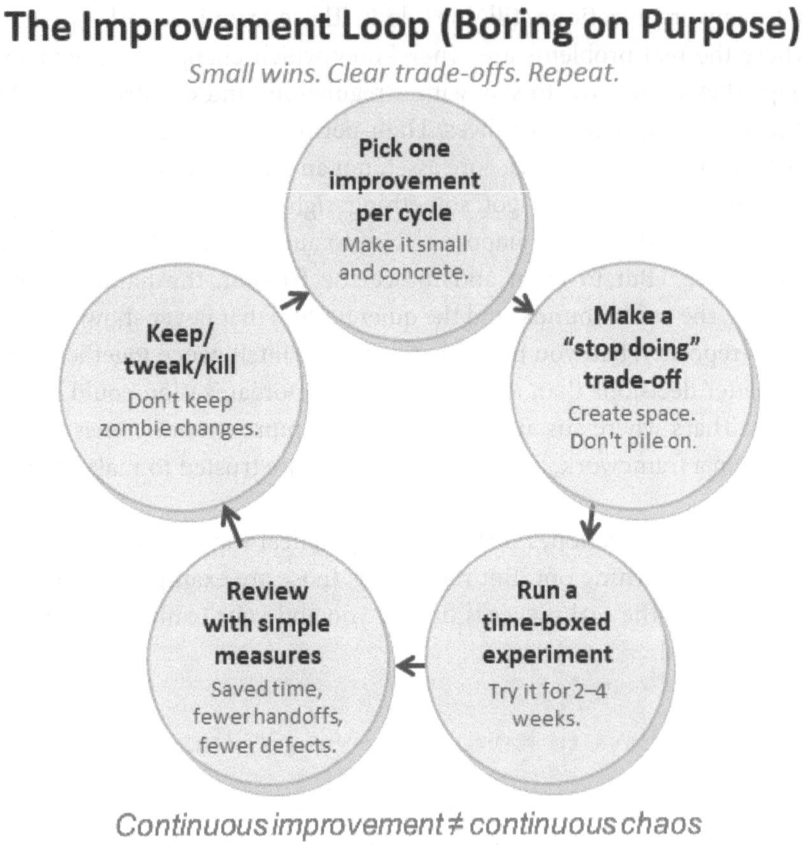

Figure 10.2 | The improvement loop

1. One Small Improvement per Cycle (Not a New Transformation Every Quarter)

You don't need a new initiative every three months. What you need is one meaningful improvement per cycle. And let's talk about "cycles" for a second. I used to get tripped up when people asked, "Do you have cycles?" or "Can we fit this into the next cycle?" I knew the dictionary definition, but in practice, they were really asking, "Do you have the capacity to do this?" A cycle is just whatever chunk of time your team uses to plan and deliver work. It doesn't have to be a sprint—it could be a week, a month, or whatever fits your context.

The point is to pick one small, concrete thing to improve each cycle. For example:

- Reduce a specific handoff that always slows things down.
- Simplify a form your team fills out constantly.
- Use automation (or AI) to handle a boring, error-prone check.

I've seen teams stuck filling out the same order forms and manually double- and triple-checking them. Over time, people get complacent, mistakes slip through, and everyone gets frustrated. That's a perfect place for a small improvement: automate the check, add validation, or change the form to make it harder to screw up. The key is to make the improvement visible: This is the improvement we're testing this cycle. Treat it like a test, not a new law. You're not saying, "This is the new way forever." You're saying, "We're going to try this and see if it helps."

At the end of the cycle, ask:

- Did this save time?
- Did it reduce friction?
- Did it decrease errors?

If yes, keep it. If not, tweak it or drop it and move on. That's how you keep improvement steady, practical, and sustainable, without overwhelming everyone.

2. Make "Stop-Doing" Lists

You can't keep piling improvement work on top of an already overloaded system and expect people not to snap. Alongside what we'll improve, you also need to ask: "What are we going to stop doing or do less of, to make space for this?" For example:

- Drop a useless report no one reads.
- Kill a recurring meeting where nothing ever changes or gets decided.
- Retire a metric that doesn't drive any action.

Sometimes you'll find something that matters only to a small group, or even just one person. That's your chance to have an honest conversation: What value are they getting from this report, metric, or meeting? Is that value aligned with what the team or organization actually needs? And could we meet that need more simply?

If the honest answer is basically, "I just like seeing this metric," or "I like this meeting because I get to talk to everyone," then it's time to negotiate. What could we do that creates real value for everyone, not just comfort for a few? This is how you keep continuous improvement from quietly turning into continuous extra work for everyone.

3. Treat Changes as Time-Boxed Experiments

Another way to avoid burnout is to stop treating every new idea like a permanent commitment. Instead, frame changes as time-boxed experiments: "We'll try this for two cycles (sprints, months, whatever fits), then review." We talk a lot about working in cycles, but let's be honest: it rarely happens consistently. Why? Sometimes teams are so busy reacting to daily fires that they never pause to reflect. Sometimes leaders expect instant results and skip the review step. Other times, no one is sure who owns the experiment—so it quietly dies in the background. If you want cycles of improvement to stick:

- Put a recurring review on the calendar. Don't just *intend* to check in—make it a habit.
- Assign someone (not necessarily a manager) to own the experiment and bring results back.
- Keep the cycle short enough for fast feedback, but long enough for a signal. Two to four weeks is a sweet spot for most teams.
- Before you start, be clear about what problem you're solving and how you'll know if it's working: saved time, fewer handoffs, fewer defects, fewer escalations, something observable.

At the end, check in:

- If it helped, keep it.
- If it sort of helped, tweak it and rerun it.
- If it didn't help, kill it and move on.

This makes change feel safer. People know they're not stuck with a bad idea forever, and it becomes easier to say, "This isn't working." That's how you build a culture where improvement is constant, but not overwhelming.[7]

4. Protect Capacity for Improvement (Don't Hide It)

If you don't explicitly make time for improvement, urgent work will eat it every single time. Set aside a small, but visible slice of capacity for improvement, say 5–10% of the team's bandwidth each cycle. Put improvement work on the same board as everything else. It's not "nice to have if we have time." It's real work with tangible outcomes.

Leaders have to respect that boundary. Don't steal that time when something squeaks. If you *do* need to use it for a genuine emergency, acknowledge it, rebalance later, and make sure improvement work isn't always the thing that gets sacrificed. This is a direct investment in productivity, engagement, and long-term return on investment (ROI). Teams that protect time for improvement see fewer recurring issues, smoother workflows, higher morale, and better retention. The payoff is real: small, regular improvements compound over time, delivering outsized benefits compared to teams stuck in constant firefighting mode.[8] This sends a clear message: improvement isn't an after-hours hobby or a "someday" project. It's part of the job. It's how things get better.

5. Let the People Doing the Work Choose What to Improve

Top-down improvements are one of the fastest ways to create burnout and cynicism. I've been there, pushing a change from above that landed

---

7 The practice of time-boxed, iterative improvement is a core principle in Agile methodologies (see Schwaber and Sutherland, *The Scrum Guide*, 2020) and continuous improvement frameworks like PDCA (Plan-Do-Check-Act). However, research shows that many organizations struggle to sustain these cycles due to a lack of ownership, inconsistent review habits, and pressure for immediate results (see Kotter, *Leading Change*, 1996; Denning, "Why Agile Transformations Fail," *Forbes*, 2018). Embedding regular, short feedback cycles and clear accountability helps overcome these barriers and makes continuous improvement a practical reality.

8 Studies show that organizations that systematically allocate time for process improvement see significant gains in productivity, quality, and employee engagement (see Womack and Jones, *Lean Thinking*, 1996; Edmondson, *Teaming: How Organizations Learn, Innovate, and Compete in the Knowledge Economy*, 2012). Protecting even a small percentage of capacity for improvement work leads to compounding benefits and a measurable return on investment over time.

like a brick in a lake. It doesn't work. Instead, ask the team: "What's one thing that, if we fixed it, would make your life noticeably easier?" Let the people closest to the work propose which bottlenecks to tackle, which tools to streamline, and which meetings to reshape or kill.

Leaders still matter here. Your job is to provide the guardrails (compliance, risk, strategy) and the air cover (removing bigger organizational blockers). But the real power comes from grounding improvement in real pain points, not leadership's latest book, conference, or shiny idea. One important caveat: team-led improvement works best when it's guided by a clear north star. Otherwise, you get local optimization: one team gets better at their corner of the world, but the overall system doesn't improve. So, give teams the freedom to choose what to improve, but anchor it in a shared goal, such as flow, customer outcomes, reliability, or sustainability.

When you empower teams to choose what to improve, you get solutions that stick, higher engagement, and a culture where people care about making things better, because the changes matter to them, not just to the org chart.

6. Treat Burnout Signals as System Data

If you've read the well-being chapter, you already know burnout isn't a personal failing. It's a system warning light. Treat those warning signs like regular health checks for your team and your workflows. Look for patterns. Are people consistently working late, not just occasionally? Are jokes about burnout becoming normal? Are mistakes, rework, or irritability creeping up? Are meetings getting quieter, with less real engagement?

When you spot those signals, don't fall back on "just push through." Use them as prompts for a system tune-up:

- What can we remove, simplify, or defer?
- Are we trying to improve too many things at once?
- Are we piling change on top of full workloads?

Make it normal to say, "This is too much," and respond by adjusting priorities, not blaming individuals. Burnout signals aren't noise. They're signals from the system, and they're telling you where to focus next.

7. Keep Your Improvement Process Boringly Simple

If your improvement process is more complicated than the work you're trying to improve, you've lost the plot. Avoid giant, continuous improvement frameworks, 20-step templates, and overengineered change programs. They don't reduce friction. They add it. Instead, keep it simple. In retros or regular team check-ins, ask:

- What slowed us down?
- What helped?
- What will we try next time?

Track a few basic before-and-after indicators: cycle time, number of handoffs, defects/rework, and escalations. The simpler your approach, the more likely people are to use it, and the more likely you'll see real, lasting improvement.

8. Normalize Saying "This Isn't Working"

Continuous improvement dies when people are afraid to admit a change was a bad idea. Make it normal and safe to say: "We tried this. It didn't help. Let's stop." Leaders should model it too: "I pushed for this change. It's not doing what we hoped. Let's adjust." This reduces sunk-cost stubbornness and keeps your system from filling up with zombie processes and rituals no one believes in. The faster you can say, "This isn't working," the quicker you can focus on what does.

9. Tie Improvements Back to Purpose, Not Just Efficiency

If improvement is only about squeezing more output from people, they'll check out. Instead, connect improvements to what actually matters: better outcomes for customers, less stress for the team, and more time for deep work, learning, or creative problem-solving. For example: "We're automating this check so you can spend more time solving problems, not copy-pasting data."

When people see improvement as about making the work and outcomes better (not just "doing more with less"), they're far more likely to get on board, and far less likely to burn out. Continuous improvement isn't about reinventing everything or launching a new initiative every quarter. It's about saving time, reducing friction, and making work and outcomes better in ways people can actually feel.

Real improvement looks like small, visible experiments, paired with a clear stop-doing list so you're not piling more onto an already overloaded system. It protects capacity for improvement, lets the people closest to the work help decide what to fix, treats burnout as system feedback, and keeps the process simple enough that people will actually use it. When you connect improvement to purpose, not just efficiency, you build a culture that gets better without burning everyone out. One caution: small improvements still need a systems view, or you'll accidentally optimize one corner while the whole system stays stuck. A team can get faster by pushing work downstream. A department can look efficient by adding approvals that slow everyone else down. Keep a bigger goal in view—end-to-end flow, customer outcomes, reliability, sustainability—and ask: where does work wait the longest? Where do handoffs create rework? What constraint is limiting flow across the whole system?

If you want improvement to stick, treat it like real work at the leadership level, too. Talk about improvement experiments in staff meetings. Celebrate the people who remove friction—not just the people who ship the most. Promote leaders who improve how work gets done—not leaders who demand more output while the system stays broken.

But there's one big enemy of sustainable improvement: chasing every new trend that pops up on LinkedIn, at conferences, or in the latest glossy consultant deck. Most organizations don't have a continuous improvement problem. They have a shiny-object problem. They jump from framework to framework, tool to tool, buzzword to buzzword, and call it progress while their culture and capability erode.

In Chapter 11, we'll explore how chasing trends burns through your time, money, and morale, kills real learning, and prevents you from refocusing on what matters.

CHAPTER 11

THE COST OF CHASING TRENDS

Shiny Objects Versus Real Agility

Many organizations don't have an agility problem. They have a trend-chasing problem. Instead of focusing on real, grounded, continuous improvement tied to outcomes, they jump from one big thing to the next. Whatever's hot this quarter. You see it everywhere: framework hopping from SAFe™ to LeSS to the Spotify model and back again; new labels like product-led, artificial intelligence (AI) first, objectives and key results (OKRs), teal, or holacracy; and tool churn that treats project platforms like a fashion show.

Trends aren't going away. People talk. Books get written. Conference talks go viral. LinkedIn fills up with must-try models. You're not going to escape that. But here's the line I want you to remember: if your strategy changes every time LinkedIn does, you don't have a plan. You have a reflex (see Figure 11.1).

Why Leaders Chase Trends (The Psychology)

I'm not here to play therapist. But with a psych degree and a career spent watching behavior, I've seen the patterns up close. And to be fair, this isn't just a leader problem. Trend-chasing is a system problem. Leaders, teams, vendors, consultants, boards, and the market all reinforce it. That said, when trend-chasing shows up at the top, it usually comes down to two big drivers: fear and ego (see Figure 11.2).

Trend-Chasing Cycle (The Reflex Loop)

If your strategy changes every time LinkedIn does, you don't have a plan. You have a reflex.

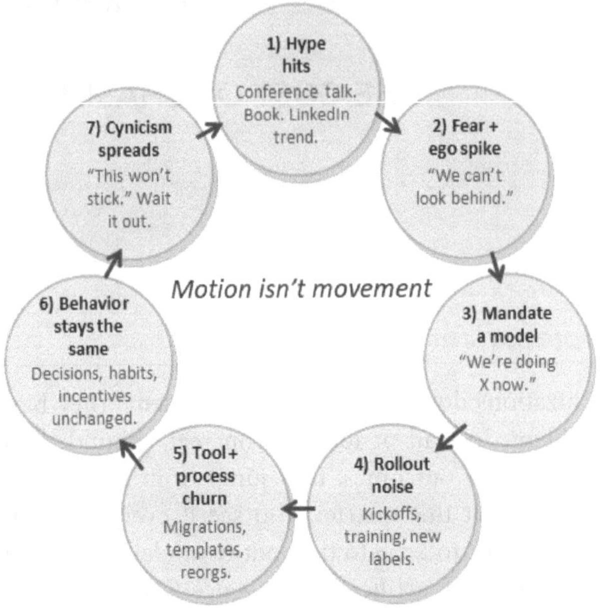

Figure 11.1 | Reflex loop

Why Leaders Chase Trends (Fear vs. Ego)

Same behavior. Different driver.

	Fear	Ego
Story	"We can't look old school."	"Look how innovative we are."
What it looks like	Copy what Company Y is doing.	New framework. New labels. Big rollout.
What it avoids	Loss of control, status, comfort.	Humility and hard self-reflection.
Result	Surface change. Same behavior.	Signaling replaces learning.

Figure 11.2 | Why leaders chase trends

Fear: "We Can't Look Old-School"

There's a real fear of being left behind. It sounds like this:

- "Everyone else is doing X. We can't look old-school."
- "Company Y uses this framework, and they're successful, so we should copy them."
- "If we don't jump on this, we'll look out of touch."

Looking at what other companies are doing can be helpful. Blindly copying them isn't a strategy. Every organization is different: different customers, different constraints, different culture, different history. If we were all clones running identical businesses, maybe you could grab the same framework off the shelf and paste it in. Even then, someone would eventually say, "This doesn't work for us," and they'd be right. Blind copying isn't agility. It's laziness dressed up as innovation.

Ego: "Look How Innovative We Are"

Then there's ego. This is the part that wants to look innovative to the board, impressive to investors, and ahead of the curve to peers. It sounds like this:

- "Look at us—we've created a new framework."
- "We're rolling out this cutting-edge model."
- "We're not like those old companies."

That's how we ended up with 30-plus Agile frameworks floating around. It's like martial arts politics. I'm a third-degree black belt in ICTF Taekwondo, and Taekwondo has multiple federations. The running joke goes like this: someone becomes a grandmaster, disagrees with the other grandmasters, and breaks off to start their own federation.[1]

---

1 Taekwondo has a long history of organizational splits and new federations forming due to differences in leadership, politics, and technical direction (e.g., ITF, WTF/WTF→WT, and various independent bodies). The "grandmaster starts a new federation" line is a common inside joke among practitioners, reflecting this pattern rather than criticizing any specific organization. See Alex Gillis, *A Killing Art: The Untold History of Tae Kwon Do* (Toronto: ECW Press, 2008), for a historical account of Taekwondo's political and organizational fragmentation.

In Agile environments, it's the same pattern. Someone becomes a big name, doesn't like how others do it, and creates their own framework. You end up with something that might work for one organization in one context, but it gets sold like a universal solution.

Even Spotify, when they shared their model, was clear: "This is what works for us. It won't work for everyone." That's humility. That's ego kept in check.[2]

The problem isn't experimentation. Experimentation is good. The problem is when ego turns "this works for our team" into "this should be the new standard for everyone."

Trends are also a way to signal modernity without doing the hard work. It's easy to say, "We're implementing OKRs," "We're moving to the Spotify model," or "We're going AI first." It sounds impressive in slides and town halls. It reassures boards and investors that you're keeping up. But if you're not changing how decisions are made, how people are treated, or how work flows, you're not modern. You're just rebranding old behavior.

It gets worse when you add in vendors and consultants selling the next big thing, influencers and gurus hyping proprietary methods, and conference talks and books that make it sound like, "If you just adopt this model, everything will click." That's how you get what I call a *conference hangover.* Someone goes to a conference or reads a book, comes back fired up, and says, "This is it. This will fix everything." Suddenly, the whole organization is supposed to pivot around a single talk or chapter.[3] No context. No experimentation. Just: "We're doing this

2 Even the much-copied "Spotify model" was originally presented as a description of how one company happened to work at a point in time, not as a universal template. Henrik Kniberg and Anders Ivarsson explicitly framed it as "a snapshot of how we work at Spotify" and cautioned others not to treat it as a prescriptive framework. See Henrik Kniberg and Anders Ivarsson, "Scaling Agile @ Spotify" (2012).

3 While "conference hangover" is a colloquial label, the pattern is well documented: organizations adopt ideas from books, talks, or gurus as short-lived "management fads" without grounding them in context, experimentation, or sustained follow-through. See Matthew Stewart, *The Management Myth: Why the Experts Keep Getting It Wrong* (New York: W. W. Norton, 2009), for a critique of management fashion cycles, and Eric Abrahamson, "Management Fashion," *Academy of Management Review* 21, no. 1 (1996): 254–285, on how popular concepts spread quickly and then fade without delivering lasting change.

now." It's not that conferences, books, or vendors are bad. The problem is when a single talk becomes a mandate instead of an input to thoughtful experimentation.

Not all trend-chasing is driven by bad intent. Sometimes it's a recruiting move. If you're trying to hire great people, you want to look like a place that's learning, modern, and investing in how work gets done. The best people often do want strong practices, good tools, and teams that take improvement seriously. Being seen as current can help you attract talent. The trap is confusing the signal with the substance. A shiny model might get attention. But what keeps great people is a place where the work is sane, leadership is honest, and improvement actually sticks.

Author Ryan Holiday shares a story about the Buddha studying under great teachers, learning everything he could, then moving on when he hit the limits of what those schools could offer. That isn't trend-chasing. That's growth. The point isn't "don't learn from others." The point is: stop copying without thinking and start building capability that fits your context.[4]

And then there's the part no one likes to talk about: real change is uncomfortable. It breaks old norms, challenges power, and forces people, especially leaders, to look in the mirror. The good news is you don't need to be perfect. You need a foundation. In most organizations, you can build a solid foundation in about a year, enough to feel the difference day to day. Work flows better. Decisions get clearer. Problems surface earlier. People stop living in a constant reactive mode.

But sustaining it is like training. You don't finish fitness. You don't complete Jiu-Jitsu. You build capability, then you keep showing up. The discipline is what makes it real. Trend-chasing offers something way more tempting: the illusion of progress without the discomfort

4 Ryan Holiday retells a traditional account of the Buddha's early training—studying meditation under teachers such as Alara Kalama and Uddaka Ramaputta, learning what they could teach, and then moving on when he recognized the limits of those schools and needed to find his own path. Holiday uses this to illustrate the difference between learning from strong teachers and blindly copying a school forever. See Ryan Holiday, *Stillness Is the Key* (London: Portfolio/Penguin, 2019), especially the section discussing the Buddha's early teachers and the limits of existing schools.

of introspection. You can announce a new framework, roll out new terminology, and buy a new tool without ever asking, "How do I need to change?" "How does our leadership behavior need to change?" "What are we willing to stop doing?" That's the real cost of chasing trends: you get motion instead of movement. You get noise instead of growth. And you avoid the one thing that actually matters: changing how people behave.

How Trend-Chasing Kills Culture

What's the real damage of chasing trends? It quietly kills your culture. First, you get whiplash: a new framework every 12–18 months, a new operating model every year, new labels and structures before the last ones even have time to take effect. From the outside, it looks like "we're evolving." From the inside, it turns into a running joke. Most employees won't say that to senior leadership, especially in places where psychological safety is already low. They'll nod in the town hall, then roll their eyes in the hallway. And here's the thing: eye-rolling isn't healthy debate. Debate is direct. Eye-rolling is passive-aggressive. It's what happens when people don't believe they can speak honestly or don't believe anyone will listen if they do.

If you want a culture that improves, you can't ignore that signal. Set the expectation: disagree out loud, respectfully. Don't perform contempt in silence. I worked with one organization where the goal was simple: get releases out on time and avoid late-stage defects. When I walked through the workflow with a team lead, and we got to dates and accountability, he said, "It doesn't matter what date we pick. It's always going to be a week or two, maybe three, after whatever we say."

When I asked why, he shrugged: "We've tried to make changes a couple of times to hit our dates. Nothing stuck. I don't expect this to make any difference either."

That's what trend-chasing does. It creates learned helplessness. People start thinking, "This will change again soon. Just wait it out," or "Don't invest too much energy, the next thing is coming." So, they disengage. Innovation and improvement dry up because why bother fixing things locally when the whole system will be dismantled again in six months?[5]

Then you get cynicism. Town halls become eye-roll central. Leadership announces "a new way of working," and if you look around, you'll see smirks and chuckles. That's a signal: "We've seen this movie before. Nothing really changes." Again, good debate is healthy. Eye-rolling isn't. If people disagree, the goal is to bring it into the open:

- What do you not believe?
- What didn't work last time?
- What would have to be true for this to actually help?

Passive aggression is what shows up when honesty doesn't feel worth it. Underneath all that, you get identity erosion. Teams lose any sense of "This is how we work well together," or "This is what we're good at." Instead, they start second-guessing themselves: "If we're changing frameworks again, we must be doing it wrong," or "This framework is 'proven' everywhere else, so the problem must be us."

Constant reinvention prevents deep mastery of anything. It stops people from learning how to adapt a system to their context. It keeps everyone in permanent beginner mode. And we rarely ask the obvious questions:

- Are we changing so much, so fast, that nothing has a chance to work?
- Are we looking at the behaviors that need to change or just swapping wrappers?

If you don't deal with those, no framework on earth is going to save you.

---

5 The pattern where people stop trying to improve because past efforts never stick closely resembles "learned helplessness" in psychology when repeated failed attempts to change a situation lead individuals to believe nothing they do will matter. See Martin E. P. Seligman, *Helplessness: On Depression, Development, and Death* (San Francisco: W. H. Freeman, 1975). In organizational settings, similar dynamics show up as change fatigue and disengagement when waves of initiatives come and go without real impact.

The Real Cost in Time and Money

Beyond the cultural damage, there are real costs in time and money associated with trend-chasing. These are the direct costs everyone expects: training people on the new framework, tool, or model; consultant and certification fees; licensing; tool migrations; and integrations across teams and environments. All of that is easy to put in a budget line: "This rollout will cost X." "This training program is Y." People see those numbers and think, "Okay, we can afford that." On paper, it looks like a reasonable investment.

What most organizations underestimate is the indirect cost. Yes, you should expect a productivity dip during change. That's normal. But when you're changing constantly, the dip never fully recovers. It becomes a staircase going down: each transformation knocks you down a step, and you never quite climb back up before the next one hits.[6]

People are constantly relearning tools and roles, attending workshops and kickoffs, but often not changing their actual behavior. The surface changes. The habits don't. Time spent on kickoff events, training sessions, and new-model briefings that don't lead to different decisions or habits is wasted capacity. That's delivery time you never get back.

How to spot the indirect cost

Signal: Lots of training, kickoffs, and new-model briefings, but day-to-day work feels the same.

What it usually means: You changed the surface (tools and labels), not the behaviors (decisions and habits).

Quick test: "What specific behavior changed after the last rollout concerning how we plan, decide, collaborate, or handle conflict?"

First move: Pick one behavior to change on purpose and make it visible for 30 days, then review.

6 Frequent, overlapping change initiatives can create a persistent drag on performance, as people spend more time adapting to new structures and tools than improving actual work. Research on "change saturation" and initiative overload shows that organizations often underestimate these indirect costs and the cumulative productivity loss. See Jeffrey M. Hiatt, *ADKAR: A Model for Change in Business, Government and Our Community* (Loveland, CO: Prosci, 2006), and Donald Sull, Charles Sull, and James Yoder, "Straight Talk About Your Organization's Culture," *MIT Sloan Management Review* 59, no. 1 (2017).

There's another cost people miss: trend-chasing makes everyone sound the same. When every vendor and every internal change effort uses the same buzzwords and promises the same "proven model," it gets harder to tell what's real and what's just packaging. Trust drops. Attention drops. People stop listening because they've heard it all before. Inside organizations, that sameness turns your message into wallpaper and your change effort into background noise. And when the work starts to feel performative instead of practical, your best people don't argue—they disengage, or they leave for somewhere that feels more honest.

Every hour you spend chasing the new thing is an hour you're not spending fixing real, known problems, improving the system you already have, or deepening capability in what's already working. That opportunity cost is expensive. It's like this: you build a house planning for white doors. During the final review, you notice the doors are brown. What's the straightforward fix? Paint the doors white. What trend-chasing organizations do is the equivalent of tearing down the whole house and rebuilding it from scratch just to get white doors.

That's what it looks like when you throw out your current way of working and import a whole new framework instead of fixing the specific problems you already know about. You're not solving. You're restarting. It makes no sense, and it's expensive. There's a real cost in time, money, and energy when you keep doing this. The bill doesn't just show up in a spreadsheet. It shows up in culture, in morale, and in the quiet decisions people make to stop caring.

The Cost in Team Morale and Capability

Trend-chasing doesn't just hit budgets and timelines. It quietly grinds people down and drains their capacity to care and perform. After a few cycles of "new framework, new model, new tool," you start to see emotional fatigue set in. People get tired of hearing, "This time it's different," or "This is the model that will finally fix everything." They've already lived through the last big change, and the one before that, and the one before that. Each time there's a big launch and a lot of noise. Then, slowly, everything drifts back to the old behavior. Eventually, it's not anger, it's exhaustion.

How to spot emotional fatigue

Signal: Flat, polite agreement in public but dark humor in private.

What it usually means: People don't believe it will stick, so they're conserving energy.

Quick test: "If we do nothing different in our day-to-day behavior, what do you think will happen?"

First move: Stop selling certainty. Name the risk out loud and commit to one behavior change you'll actually protect for 30 days.

Meanwhile, skills erode where they actually matter. Instead of deepening expertise in product thinking, engineering, collaboration, and honest feedback, people become experts in surviving reorgs, translating buzzwords, and mapping old roles to new roles just to keep doing the same work under a different label. You end up with teams that are shallow in five frameworks, deep in none, and very good at enduring change, but not necessarily at doing great work.

How to spot skill erosion

Signal: People can explain the new labels but can't name what they've genuinely improved at.

What it usually means: The organization is investing in terminology, not capability.

Quick test: "What have we gotten better at in the last 12–18 months? Craft, delivery, collaboration, or just navigating structure?"

First move: Pick one capability to deepen feedback, discovery, quality, or flow, and invest in it long enough to see results.

Constant ramp-ups and ramp-downs take a toll. There are always new processes to learn, new expectations to meet, and old work that still needs to get done. You're always in transition, never in stable flow. Burnout doesn't always show up as people collapsing. It often shows up as numbness, minimal effort, and the quiet attitude of, "Just tell me what you want, and I'll do the bare minimum."

How to spot burnout

Signal: More mistakes, more rework, more irritability, and more people going quiet in meetings.

What it usually means: Discretionary effort is gone. The system has worn people down.

Quick test: "What are we asking people to absorb on top of a full workload?"

First move: Reduce load (WIP, meetings, initiatives) before you add another "improvement."

This all comes to a head when the people you can least afford to lose exit the premises. Those who care most and want to build something meaningful are often the first to leave. They don't want to spend their careers repainting the same room a different color every year while nothing fundamental changes. They want clear direction, a stable enough environment to get good at something, and the ability to actually ship and improve things. If they can't get that where they are, they'll go somewhere else.

How to spot talent loss

Signal: Your most engaged, improvement-minded people are leaving first.

What it usually means: You're not just losing headcount—you're bleeding capability.

Quick test: "Who are we losing, and what do they have in common?"

First move: Ask your top performers what would make them stay, then immediately remove one central friction point.

Trend-Chasing Inside Agile Itself

This isn't just a business-in-general problem. It's alive and well inside the Agile space too. You see it in framework hopping:

- "We tried Scrum. It didn't work."
- "We're doing SAFe™ now."
- "That didn't work either, so we're going to the Spotify model."
- "Now we're creating our own hybrid."

None of these gets enough time to settle, enough time to adapt to the actual context, or enough real behavioral change at the leadership level. It's like changing your workout plan every week and then complaining you're not getting stronger.

You also see it in ritual swapping. Teams switch estimation methods, change their retros every sprint, and rotate metrics every quarter. On their own, none of those things is bad. Experimenting with practices is healthy. But if you're constantly swapping them without understanding why, measuring impact, or letting people get good at anything, you're not doing agility. You're doing Agile cosplay. It looks like agility from a distance. Up close, it's just costume changes.

How to spot ritual swapping

Signal: Your team can list more frameworks you've tried than behaviors that have actually changed.

What it usually means: You're rotating wrappers (frameworks and rituals), not changing decisions, conflict habits, or leadership behavior.

Quick test: "Name three concrete behaviors that changed in how we plan, decide, and handle conflict."

First move: Pick one behavior to change on purpose, protect it for a full cycle, and measure whether it reduced friction or improved outcomes.

Red Flags: Are You Chasing Trends?

Let's make this practical. Trend-chasing has tells. Ask yourself:

- Do we change frameworks more often than we change leadership behaviors?
- Have we switched core tools multiple times without a clear, measurable benefit?
- Do we launch more initiatives than we can name real improvements from?

Listen to the language:

- "We're moving to X model now."
- "This framework will finally give us consistency."
- "We're rebranding our teams again."

None of those lines is evil on its own. But if that's most of the conversation, you're probably living in shiny-object land. Then ask the questions that expose reality:

- Why are we doing this (without reading a slide)?
- What problem does this solve for us, in our context?
- What will we stop doing to make room for it?
- What will look different in 6–12 months if it worked?

If you can't get straight answers, you're not improving. You're burning energy chasing something that keeps moving while you stay stuck.

How to Focus on What Actually Matters

The antidote to trend-chasing isn't to never change. It's changing with purpose. That starts with problems, not trends. Instead of asking, "What's the latest model everyone's using?" start with a few grounded questions: What hurts the most right now? Where are we wasting the most time, money, or energy? What keeps getting in the way of delivering value?

Once you're clear on the real pain points, run any change through a behavior-first filter. Don't start with slides or structures. Start with people. Ask: What behaviors would need to change for this to work? Are we actually willing to change those behaviors? And specifically, what would leaders need to do differently? If you can't answer those, you're not ready for the change. You're just shopping for a new label. From there, slow down to go faster. Instead of stacking five big bets on top of each other, commit to one meaningful change at a time. Give it time to stabilize. Let people learn, adapt, and improve it. Depth beats constant reinvention every time.

Finally, anchor every improvement in something that matters: better outcomes for customers, better flow of work (less friction, fewer handoffs, clearer decisions), or better well-being (less chaos, more clarity). If a proposed change doesn't connect to at least one of those, then the honest question is: why are we doing this at all?

Guardrails for Evaluating New Trends

You don't have to be the grumpy "we never change anything" person to avoid trend-chasing. You need guardrails so you don't chase every shiny thing that comes along. Start here:

- What problem does this solve for us, specifically? Not in the book. Not in that case study. For us. In our context. With our constraints.
- What will we stop doing if we adopt this? If the answer is "nothing," you're not changing. You're just adding complexity. Real change displaces something.
- What behaviors would leaders have to change? If the answer is "none," it's probably framework performance. No leadership change, no transformation.
- How will we measure whether it's helping? What will look different in three, six, or twelve months, in outcomes, flow, or well-being?

And here's the execution rule that saves your sanity: pilot, don't roll out. Try it with one team or area first. Learn from it. Adapt it. Only then decide if it's worth scaling. That's how you stay open to new ideas without sacrificing your culture, your people, or your sanity.

Chasing every new framework, model, and buzzword doesn't make an organization more Agile. It makes it tired, cynical, and shallow. It creates cultural whiplash, learned helplessness, and a revolving door of "this time it's different" initiatives that never touch real behavior. It burns time and money on training, tooling, and reorgs, while obvious, fixable problems sit right in front of people. Worst of all, it erodes capability: teams get good at surviving change, not doing great work. This doesn't have to be the pattern. Organizations can choose something better. They can stop reacting to every new trend and start paying attention to what actually helps: clearer decisions, healthier teams, better flow of work, and leadership behavior that people can trust.

That's where real agility starts. Not in a new label. Not in a glossy rollout deck. In people who are willing to slow down, tell the truth about what isn't working, and improve things in ways that actually last. If we can do that, the future of agility gets a lot more human. It stops being about optics, frameworks, and status, and starts being about how we work together under pressure without losing ourselves in the process.

In Chapter 12, we'll look at the future of agility. What it could look like if we drop the nonsense, put well-being and behavior at the center, and choose to lead differently. Not perfectly. Just honestly, consistently, and in a way that gives people room to do great work and still be human at the end of the day.

CHAPTER 12

THE FUTURE OF AGILITY (IF WE BUILD SOMETHING THAT LASTS)

What This Future Is

Up to now, we've stripped away a lot of nonsense. We've looked at what's performative, what burns people out, and what keeps organizations stuck chasing labels instead of building anything real. We've also looked at a better path: environments where people can do good work, learn, adapt, and grow without getting crushed in the process. Underneath all of this, regardless of industry, company size, or context, there's usually one thing people want: to build something that matters. Something that outlasts them. Something worth being part of. That's what this chapter is about.

This isn't a prediction, and it's not a new framework. I don't have a crystal ball, and I'm not interested in handing you another branded model to worship. This is a vision of what becomes possible if we stop chasing novelty and start focusing on what actually matters: people, behavior, and the conditions that help teams do great work. If we don't deal with people and behavior, nothing else will stick.

The funny part is the Agile Manifesto already pointed us in this direction: *individuals and interactions over processes and tools.* They knew then what a lot of organizations still struggle to admit now. If you help people understand the change, involve them in it, and make work better instead of harder, they'll move with you. The future of agility isn't a new framework or a shinier model. It's a different way of showing up (see Figure 12.1).

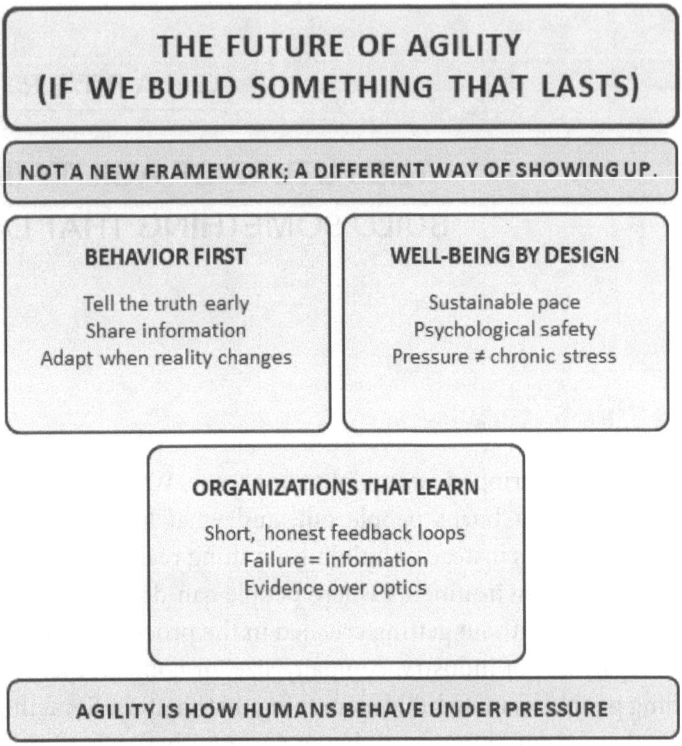

Figure 12.1 | The future of agility

Here's the Future Worth Building

1. Work That Doesn't Break People

If we build something people can actually live inside, we can create work environments that don't chew people up and spit them out. Imagine teams delivering real value and still going home with energy left. Priorities are clearer. Not everything is "urgent" all the time. And when something really is urgent, people don't pretend otherwise. They talk honestly about trade-offs instead of quietly dumping the cost onto whoever has the least power to push back. That means asking better questions before the panic takes over:

- What are we trading off?
- What breaks if we say yes?

- What breaks if we say no?
- What can we reduce, delay, or drop to make this sustainable?

In healthy environments, people can still say "no," "not yet," or "yes, and..." without fear of punishment. That's what real empowerment looks like. Teams understand the goals, the constraints, and the consequences. They're trusted to think, not just comply. They're not expected to burn themselves out just to hit fake deadlines or protect someone else's optics. Leaders need to stop confusing commitment with self-destruction. You want to be legendary, not just a short-term hero. A hero leader jumps in, saves the day, removes the blocker, works late, and takes the hit. People remember them for a while. Their name ends up in a few stories.

A legendary leader does something harder. They teach people how to solve problems without making them feel small. They help people become more capable, more confident, and more self-sufficient. They don't just fix the issue in front of them. They leave behind better thinking. I've had a few legendary leaders in my life. I'll talk about them until the day I die, because they didn't just solve problems for me. They taught me how to think, how to learn, and how to help other people do the same. That's the kind of leadership the future of agility needs (see Figure 12.2).

In that future, well-being isn't a perk or a recovery plan. It's a prerequisite for performance. It's not "we'll pay for your yoga app" while the work itself stays broken. It's designing work so people don't have to destroy themselves to succeed. It's caring about how work feels, not just what people do after hours to recover from it. That kind of environment doesn't happen by accident. But it also doesn't require magic. It takes intention, honesty, and the courage to stop treating exhaustion like proof that people care.

2. Organizations That Actually Learn

If we drop the noise, we can build organizations that actually learn—not just talk about learning. In a real learning organization, feedback loops are short, honest, and used. We don't collect lessons learned and file them away like a compliance artifact. When something fails, we don't treat it like a character flaw. We treat it like information.

Hero Leader vs. Legendary Leader

One fixes problems. The other builds problem solvers.

	Hero Leader	Legendary Leader
Default move	Jumps in Takes over	Coaches Creates clarity
In a crisis	Becomes the bottleneck Works late	Supports the team; Let's them solve it
What it teaches	"Wait for the boss"	"We can handle this"
Long-term effect	Dependency grows Burnout risk rises	Capability grows Calm spreads
Legacy	Remembered for saves	Remembered for growth

Build leaders, not followers.

Figure 12.2 | Hero versus legendary

- What happened?
- Why did it happen?
- What are we going to do differently next time?

That also means dropping a lot of bad habits: no scapegoat hunts, no shaming the person closest to the problem, and no "because I said so" decisions dressed up as leadership. Evidence matters. Context matters. Patterns over time matter more than hierarchy or hunches. This is what Peter Senge was getting at in *The Fifth Discipline*. Learning organizations don't just react to problems. They use problems to understand the system, challenge old assumptions, and improve how they operate. They treat learning as a discipline, not a slogan.

In that kind of environment, teams are trusted to experiment, adapt, and improve without having to go through a ridiculous approval chain for every small change. Experiments are time-boxed. Results get reviewed. What works gets kept. What doesn't gets changed or dropped. Leaders change their minds when the evidence changes. That's real agility! And it matters because learning organizations don't just make faster decisions. They make better ones. They waste less time defending bad calls. They spend less energy pretending. And over time, they get better at responding to reality without turning every surprise into a crisis.

3. Customers Who Feel the Difference

If we get this right, customers feel it too. They experience products and services that solve real problems and improve based on actual feedback, not internal politics or leadership's latest obsession. They don't feel like beta testers for your internal churn. They don't get dragged through the side effects of your organizational confusion. Of course, customers will still see rough edges sometimes. That's normal. But they shouldn't feel like they're living inside your chaos. What customers want is pretty simple:

- **Consistency**: Things work the way they expect, most of the time.
- **Clarity**: They understand what's changing and why.
- **Responsiveness**: When something's off, you listen and adjust.

When teams are empowered, leaders build trust instead of noise, and well-being is treated as part of good system design. Not because you told them about your model. Because you became more reliable. Customers don't care what you call your framework. They care whether they can count on you.

What Makes That Future Possible

If you've made it this far, you already know where I stand. The future I'm talking about is behavior-first and well-being focused. Not framework first. Not certification first. And "Agile" is an adjective, not an identity, a job title, or a personality type. Frameworks don't make organizations adaptive. People do. Tools don't create trust. Behavior does. Certifications don't make teams healthier, braver, or more honest under pressure. Daily choices do. If we're serious about building an environment that holds up under pressure, there are a few behaviors that need to become normal.

1. Tell the truth, even when it's uncomfortable.

I've worked in organizations where I could walk up to a leader and say, "This isn't working," "We're going to miss this date," or even, "Three-quarters of a database got wiped out, and it was my fault." And I didn't get humiliated for it. I got help. We dealt with the problem, learned from it, and moved on. That's the behavior we need more of—truth over comfort. Reality over optics. If people can't tell the truth about what's happening, everything else becomes performance. Status updates become fiction. Risks get hidden. Problems surface late. And by the time reality finally shows up, it shows up with interest.

2. Share information instead of hoarding it.

It's still surprisingly common for people to hoard information because they think it protects their value. Sometimes that's explicit knowledge: documents, checklists, data, processes, how-tos. Other times it's tacit knowledge: the unwritten stuff, the instincts, the "here's how this really works around here" understanding that only comes from lived experience. The old logic says, "If I share everything I know, I'll lose my edge." That's nonsense.

Sharing what you know doesn't make you replaceable. It makes the team stronger. It reduces single points of failure. It helps people adapt faster, avoid avoidable mistakes, and build on each other's thinking instead of protecting tiny islands of expertise. Real agility depends on collective intelligence. And collective intelligence only grows when people share both the documented stuff and the lived stuff.

3. Ask for help before things explode.

I've seen this pattern too many times: things are going sideways, everyone knows it on some level, but nobody says it clearly until the whole thing blows up. Daily scrums become status performances. People say, "I'm still working on X," when what they really mean is, "I'm stuck,"

"I'm behind," or "I have no idea how to solve this." Now we're building artificial intelligence (AI) tools to detect when things are going wrong and surface them in stand-ups. AI can help spot patterns. But it still doesn't know what's happening in someone's head. We're humans. We can communicate with each other.

In a behavior-first environment, asking for help isn't treated like weakness or failure. It's treated like good system behavior. It helps the team respond earlier, reduce damage, and solve problems while they're still manageable.

4. Adapt decisions when reality changes.

Plans matter. But reality matters more. If a bridge is out, you don't keep marching toward where the bridge used to be because that's what the map says. You change course. Yet organizations do this all the time. Reality changes. Constraints shift. New information shows up. The original plan stops making sense. But people keep pushing forward because changing direction feels embarrassing, political, or inconvenient.

Behavior-first agility means we notice when reality shifts, and we adapt. We don't treat plans as sacred. We treat them as useful until they stop being useful. That's not chaos. That's maturity. And through all of this, frameworks stay where they belong: as tools, not religions. You adapt them to your context. You don't copy-paste from case studies and expect your organization to magically become Spotify, Netflix, or whoever people are worshipping this year. You're you. Act like it.

In this kind of future, well-being isn't an afterthought. It's part of the design. Work is structured around a sustainable pace. People can do their jobs within something resembling a normal workday. Boundaries are clearer. "Always on" heroics stop being the default. Psychological safety isn't optional. You can't claim to care about well-being if people are afraid to speak up, if saying "no" is career-limiting, or if every week feels like a death march.

Leaders need to understand the difference between pressure and stress. Pressure can be healthy. It can stretch people, sharpen thinking, and create growth. Stress is different. Stress is overload. It's chronic anxiety, fear, exhaustion, and the feeling that no matter how hard you work, it still won't be enough. If you've lived in a high-stress environment for a long time, your sense of normal gets warped. You stop noticing how bad it is because bad becomes familiar.

So how do you find the line? You talk to people. You ask what feels heavy right now. You ask where the work is grinding them down. You ask what's happening outside of work that might be affecting capacity. You're not their therapist, but you are still dealing with human beings, not interchangeable resources. When you do that honestly, people will usually tell you more than you expect. They'll show you where the line is between healthy stretch and destructive overload. Your job is to build a stretch-don't-snap environment: enough challenge for people to grow, not so much that they break. That's good system design.

People should be able to do a solid day's work, feel like they accomplished something meaningful, and still have energy left for the rest of their lives. And yes, you can measure this. Not only just output and velocity but also signals of team health and sustainability: perceived workload, ability to focus, frequency of fire drills, burnout risk, quality of collaboration, and whether people can recover without disappearing for a week. Those signals are real. In many cases, they matter more than the neat little delivery metrics leaders like to put on slides.

If this future is going to last, leadership has to change, too. Not just the language. Not just the slide deck. The actual behavior. In this kind of environment, leadership isn't reserved for people with big titles or corner offices. Leadership is about helping people do better work, navigate reality, and grow stronger over time. That means leaders remove obstacles rather than create them. They create clarity instead of noise. They model vulnerability instead of fake toughness. They support people rather than micromanage them. They respond to what's actually happening, not just what the plan said was supposed to happen.

And yes, some things will always stay confidential. Mergers, legal issues, sensitive financials, fine. Nobody is asking leaders to spill every secret. But they can still be honest about what they can share, what they can't, and what's true right now. They can still treat people like adults. That matters more than many leaders realize.

The system doesn't just run on process. It runs on what gets rewarded. If you reward heroics, politics, and individual output at the expense of everyone else, you will get more heroics, more politics, and more burnout. If you say you want teamwork but only promote the loudest individual performers, people will notice. And they'll adapt accordingly. So if we want healthy, team-based delivery and healthy individual growth, we need to reward different things. We need progression paths for people who coach, unblock, teach, improve flow, share knowledge, and make better decisions possible. We need to recognize the people who make the system stronger, not just the people who look impressive inside it. That includes Scrum Masters, coaches, staff-level contributors, team leads, and anyone else whose real value comes from helping other people succeed.

People should be able to build a career by improving the environment around them, helping other people get better, and making work more effective without making it more miserable. That's the kind of system worth building—ecosystems, not feature factories. And ecosystems don't thrive on command-and-control. They thrive on clarity, trust, support, and adaptation. Leaders at every level need to get comfortable saying things like:

- Here's where we're going.
- Here's why it matters.
- Here's what's nonnegotiable.
- Here's how I'll support you while you figure out the best way to get us there.

That's not a framework. That's a different standard of leadership under pressure.

What This Looks Like Day to Day

A lot of visions sound good until you ask the obvious question: Okay, but what would this actually feel like in real life? If we built this kind of environment, you wouldn't just see it in strategy decks or leadership language. You'd feel it in the day-to-day work.

For starters, you'd see fewer fire drills and fewer war rooms. You know the kind I mean. Something breaks, and suddenly everyone and their uncle gets dragged into a call. Senior leaders. Managers. People who have no business being there. Half the organization is staring at the same screen, all trying to look useful while a handful of people do the actual work. In a healthier system, self-managing teams handle most of that. They have the trust, skill, and authority to respond quickly, keep the right people informed, and deal with the root cause without turning every issue into a full-company crisis.

You'd also see fewer fake deadlines. The classic version is a project that's "definitely" going to be done in months, and years later, it's still limping along. I once got pulled into a project with an eight-month schedule. It ended up taking two years and four months. That kind of thing doesn't happen because people are lazy. It happens because organizations confuse declarations with reality. They pick a date that sounds good, then pressure people to make the plan fit the fantasy.

In a sane environment, teams and leaders work with reality rather than against it. What can we actually deliver? By when? What trade-offs come with moving faster? What gets riskier if we compress the timeline? What needs to move if this becomes the priority? That doesn't make organizations slower. It makes them less delusional.

You'd also see fewer useless meetings. The whole "this meeting could have been an email" joke stops being funny when it's your actual life. Too many teams spend huge chunks of their week trapped in status meetings, update meetings, pre-meetings, and meetings about meetings. In a healthier environment, meetings are for real dialogue, decisions, and collaboration. Not for reading updates out loud to people who could have read them faster on their own.

That shift creates room for something most organizations say they value but rarely protect: deep work. People get more uninterrupted time to think, design, solve problems, and improve the system itself. They're not constantly context-switching between Slack pings, calendar invites, and artificial urgency. Focus starts to matter again. And when people have time to think, the quality of the work usually goes up.

You'd also see more real collaboration. Not just people sitting on the same call or sharing the same board. Real collaboration means shared understanding. It means people know how work moves across roles, where handoffs create friction, and how to pass context without tossing problems over the wall. It also means decisions get clearer. Teams understand that trade-offs are real. If we do X, we probably can't do Y right now. If we pull this forward, something else moves back. If we want more quality, speed may change. If we want more speed, risk may change.

That kind of clarity sounds basic, but many organizations avoid it because trade-offs make people uncomfortable. They'd rather pretend everything can be top priority at once. It can't. In a healthier system, teams are trusted to make many of those calls locally. They don't need every decision to crawl up three layers of hierarchy before anyone can move. They have shared goals, shared context, and enough trust to respond to reality together. That's what mature autonomy looks like.

This kind of future changes careers, too. People don't have to destroy themselves to be seen as high performers. They can grow as craftspeople, as leaders, and as human beings without turning burnout into a badge of honor. Yes, individuals can still perform at a high level. But the bigger win is being part of a high-performing team—one where people support each other, improve together, and feel proud of the work without having to fight each other for oxygen. That changes what Agile roles become. You get fewer framework police and more behavior and systems coaches. Agile coaches, Scrum Masters, and leaders in this space stop acting like process enforcers and start acting like people who can read behavior, understand team dynamics, spot system friction, and help teams and leaders improve the environment around the work. That's a much more useful job. You're not there to force the ecosystem into a template. You're there to help it grow in a healthier direction.

And yes, this changes how organizations compete. In this kind of environment, your edge doesn't come from having the newest framework, the prettiest slide deck, or the trendiest language. It comes from things that look less exciting in a keynote and matter a lot more in real life:

- How quickly you adapt when reality changes.
- How much you trust teams to make decisions and own outcomes.
- How fast you can learn from what's happening and respond.

That's it—adaptability, trust, and learning speed. Those things are not soft. They are not side benefits. In a market economy, they are competitive advantages.

Organizations that can sense change early, make better decisions faster, and improve their own systems without burning out their people will outperform organizations that keep chasing the next shiny thing. That's the future of agility if we build something that lasts. Less framework performance. More honest, human, behavior-first work. Organizations that don't just survive change, but grow through it, without breaking the people inside.

Here's Your Part in Building It

It would be nice if this kind of future just showed up on its own. It won't. It gets built one decision, one conversation, one boundary, one act of honesty at a time. Not through a rebrand. Not through a new tool rollout. Not through another expensive transformation plan that says all the right words while people keep drowning underneath it. If we want a more human, useful, sustainable version of agility, then all of us have a part to play.

If you're a leader, your job is not to perform certainty. Your job is to create clarity, make trade-offs visible, tell the truth about what's happening, and stop asking people to absorb the cost of bad decisions in silence. That means being honest when priorities conflict. It means admitting when a date is fantasy. It means asking whether the system is helping people do good work or just making them look busy. It means rewarding learning, collaboration, and good judgment—not just heroics and optics. It also means looking in the mirror.

- Are you creating the kind of environment where people can speak honestly?
- Can people tell you bad news without bracing for impact?
- Do your teams know what matters most right now?
- Are you building capability in others, or are you making yourself the center of every solution?

You don't need to be perfect. But you do need to be real. People build their understanding of work around what leaders tolerate, reward, and model. If you say you care about well-being, trust, and adaptability, but your behavior says speed at all costs, people will believe your behavior. They should.

If you're a coach, Scrum Master, or change practitioner, your part is bigger than protecting the process. You are not here to defend a framework like it's sacred text. You are not here to win arguments about whether a team is doing Scrum properly. You are not here to become the local guardian of events, templates, and vocabulary while the actual system keeps hurting people. Your job is to help people see what's really happening.

That means noticing patterns. Naming friction. Making work visible. Helping teams tell the truth. Helping leaders understand the impact of their decisions. Protecting space for learning. Challenging nonsense when you see it. Translating complexity into language that people can actually use. And yes, sometimes that means walking into uncomfortable conversations.

- Sometimes it means saying, "This isn't a team problem. This is a leadership problem."
- Sometimes it means saying, "This process is creating more noise than value."
- Sometimes it means saying, "We keep talking about agility while rewarding the exact behaviors that make it impossible."

That's the work. Not framework performance. Not role protection. Not looking useful in a quarterly transformation update. Real help.

And if you're someone living inside these systems every day, your part matters too. You may not control the org chart. You may not control the budget. You may not get to rewrite the strategy. But you still shape the environment around you more than you think:

- You can tell the truth.
- You can ask better questions.
- You can ask for help earlier.

- You can share what you know.
- You can stop pretending everything is fine when it isn't.
- You can refuse to confuse performative busyness with meaningful work.

You can also pay attention to how work feels. Not in a vague, fluffy way, but in a practical way.

- Where is the friction?
- Where are people afraid?
- Where do decisions stall?
- Where does urgency keep getting manufactured?
- Where are people carrying too much for too long?

Those questions matter because systems don't become healthier through intention alone. They get healthier when people notice what's happening and respond to it honestly. And no, you won't always have the power to fix everything, but you can still influence the next conversation. The next handoff. The next meeting. The next moment, when someone has a choice between pretending and telling the truth.

A lot of people read books like this and immediately think they need a massive overhaul. You don't. You do not need to go back to work tomorrow and announce a revolution. Start smaller than that:

- Pick one place where the work feels heavier than it should.
- Pick one meeting that keeps wasting everyone's time.
- Pick one decision pattern that creates confusion.
- Pick one team habit that looks normal but quietly drains people.
- Pick one truth that everyone knows, and nobody is saying clearly.

The future worth building doesn't arrive all at once. It shows up in small, repeated choices that make work more honest, more sustainable, and more human. Those choices eventually compound into success.

Final Thoughts: Build Something Worth Leaving Behind

There's a line in *Rocky Balboa* that has stuck with me for years: "It ain't about how hard you hit. It's about how hard you can get hit and keep moving forward." A lot of people hear that and think it's just about toughness. It's not—not the useful kind, anyway.

The point isn't to glorify getting punched in the face by work over and over again. The point is that life changes. Work changes. Markets change. Teams change. Plans fall apart. Pressure shows up. Reality refuses to care about your slide deck. What matters is how you respond when that happens.

- Do you double down on control?
- Do you hide behind process?
- Do you protect appearances while people quietly burn out?
- Or do you adapt, learn, tell the truth, and keep moving in a way that actually helps the humans involved?

That's what this book has really been about. Not whether Agile is good or bad. Not whether Scrum is pure enough. Not whether your organization has the right roles, labels, or rituals. It's about whether we're building systems of work that help people think clearly, work honestly, learn quickly, and stay human under pressure.

If your version of agility creates fear, confusion, exhaustion, and performative compliance, then it's not helping. I don't care how many sticky notes, dashboards, certifications, or transformation roadmaps are attached to it. If it burns people out while pretending to improve delivery, it's broken. And if it's broken, we should stop defending it.

We should build something better. Something more honest. Something more useful. Something people can actually live inside. That doesn't mean throwing away every practice, tool, or framework. Some of them are helpful. Some of them are genuinely useful in the right context. But they should stay in their place: as tools in service of people, not people in service of tools.

That's the shift:

- Worship → usefulness
- Performance → practice
- Control → learning
- Burnout → sustainability

The real future of agility is not a movement with better branding, a cleaner certification ladder, or a shinier version of the same old nonsense. It is just people doing the hard, brave work of building environments where good work can happen without crushing the people doing it. That kind of work won't always be loud. It won't always look impressive from the outside. Sometimes it'll just look like a leader telling the truth, a team asking for help sooner, a coach naming the real issue, a manager protecting focus instead of creating noise, or a person deciding not to pass stress downhill.

These small things shape culture. They shape trust. They shape whether people grow or shut down. They shape whether organizations become places where people can do meaningful work or just survive long enough to update LinkedIn.

So if you take anything from this book, let it be this:

Agility was never supposed to be a brand people perform. It was supposed to be a better way for humans to work together when things are hard.

If we remember that, we might still build something worth leaving behind. And if we do it well, maybe the people who come after us won't have to spend so much time recovering from the systems we built. Maybe they'll get to inherit something better. That's the future worth building.

CONCLUSION

If you've made it this far, thank you. Really. You gave me your time. That matters to me. You stuck with me through the frustration, the pushback, the stories, the hard truths, and probably a few moments where you thought, "Yeah, exactly," or maybe, "Damn, that's a bit much." Fair enough. I can live with that.

I didn't write this book to sound clever. I didn't write it to build a brand, win points online, or take cheap shots at Agile because it makes for a catchy title. I wrote it because I'm tired. Tired of watching good people get chewed up by systems that were sold as the answer. Tired of watching teams do everything they were told to do and still get blamed when things go sideways. Tired of watching leaders duck the hard parts, then hide behind process, roles, and polished language, like that somehow makes the damage less real. It doesn't. And if I sound a bit angry about that, it's because I am.

It's not loud, performative anger. Not outrage for effect. I mean the kind that builds when you've seen the same pattern too many times, and you know people are paying for it with more than just missed deadlines or bad deliverables. They're paying with energy. With confidence. With patience. With health. With time they should've had for their family, their friends, their training, their rest, or just their own damn life. That part gets to me because work matters. It takes up too much of our lives not to. And if we're going to ask people to bring their brains, effort, care, and time into a system, then we have a responsibility to make sure that system doesn't slowly grind them down and call it improvement.

That's why this book is personal for me. Not because I think I've got it all figured out. I don't. Not because I've always handled this perfectly. I haven't. I've missed things. I've held back when I should've pushed harder. I've seen risk and not made it visible enough. I've known something was off and not had the words, leverage, or guts to say it as clearly as I should have. I've walked away from some situations knowing I could've done better. I carry that with me. I think I should. If this work means anything, it has to stay personal. It has to keep costing us something when we miss. Otherwise, it turns into another performance. Another set of nice words. Another polished message about people, while people keep paying the price.

I'm not interested in that. I care about whether people can breathe at work. I care about whether a team can tell the truth without getting punished for it. I care about whether leaders are willing to act like leaders when it's uncomfortable, not just when it makes them look competent. I care about whether people still have something left at the end of the day. That's not soft. That's not a side issue. That is the issue.

When people have nothing left, everything gets worse. Thinking gets worse. Decisions get worse. Relationships get worse. Learning gets worse. Work gets smaller, meaner, and more mechanical. People stop speaking honestly. They protect themselves. They go quiet. They do what's safest. And then everyone acts surprised when the culture feels dead.

So no, this book was never really about Agile as a brand. It was about people. It was about behavior. It was about what happens in the room when pressure shows up. That's where the truth lives:

- You can say you value transparency, but what do you do when someone brings you bad news?
- You can say you care about learning, but what happens when something fails?
- You can say you trust your team, but what do you do when they make a call you wouldn't have made?
- You can say you want agility, but if people can't speak honestly, can't make decisions, can't adapt without fear, and can't work without being drained dry, then what exactly are you building?

Those are the questions I keep coming back to. Not what framework are you using. Not what certification someone has. Not whether the board looks tidy. What are you actually building? If the answer is a system where people are overloaded, second-guessing themselves, hiding problems, and performing progress instead of making it, then I don't care how modern it looks. It's broken. And I believe a lot of people know that. They feel it. They've just been taught not to say it out loud.

Maybe that's part of why I wrote this, too. To say it out loud. To name the thing people keep dancing around. To say that a lot of what gets sold as transformation is just better-packaged confusion. To say that process is not leadership. To say that jargon is not clarity. To say that making people more exhausted is not improvement. To say that if your way of working burns people out, it is not working.

I want better than that. Not perfect. Better. More honest. More useful. More human. I want leaders who can hear hard things without punishing the person who said them. I want teams that can raise a problem before it becomes a disaster. I want workplaces where people don't have to choose between doing good work and protecting their health. I want us to stop confusing compliance with growth and stop acting like one more model, one more role, or one more workshop is going to save us from behavior we refuse to deal with. It won't.

At some point, we have to look at ourselves. At how we lead. At what we reward. At what we tolerate. At what we avoid. That's where change starts. Not in the branding. Not in the rollout deck. Not in the big announcement. In the choices people make when things get hard.

That's why I keep coming back to this: every outcome is feedback. I'd rather win, obviously. I'm not interested in romanticizing failure. Losing isn't noble. Pain isn't the point. But if we're honest, if we're paying attention, if we stop wasting energy pretending, then even the misses can teach us something worth carrying forward. That's how better gets built. Not all at once. Not cleanly. Not with some perfect plan. It gets built in real moments: awkward conversations, clearer decisions, boundaries that should've been there sooner, and someone finally saying, "This isn't working," with a room mature enough to hear it.

If this book has done anything useful, I hope it didn't just give you language. I hope it hit a nerve. I hope it gave shape to something you've felt for a while but maybe haven't known how to say. I hope it made you feel less alone if you've been sitting in systems that look fine from the outside and feel awful on the inside.

And I hope it makes you do something. Not everything. Something. Because that's how this changes: one honest move at a time.

- If you're a *leader*, change one behavior your team has been paying for.
- If you're a *coach*, stop hiding behind language that sounds smart but changes nothing.
- If you're *part of a team*, don't underestimate what can happen when someone names what's true and asks a better question.

You do not need a perfect plan. You do not need permission. You do not need another framework to hide behind. You need courage. You need honesty. You need to care enough to make the next better choice. Then the next one after that.

That's it. That's the work. That's how trust starts. That's how culture changes. That's how people start to believe work can feel different. If enough of us do that, maybe we can finally stop treating agility like a product to buy, a label to wear, or a script to follow, and start treating it like what it should've been all along: a way of working that helps people think, adapt, tell the truth, and do good work without getting crushed by the system around them.

That's what I care about and what I want to help build. If you've read this far, I'm guessing some part of you cares about that too. So start where you are:

1. Tell the truth
2. Protect people
3. Make the next better choice
4. Then do it again

That's enough to begin.

Appendix A

THE DECODER RING

Appendix A is here for one reason: to help you translate Agile-sounding language into plain reality. Most of the damage in this space doesn't come from bad intent. It comes from vague words that let people avoid hard decisions. If you can translate the language, you can surface the real problem faster, and you can stop wasting time arguing about labels. Use this as a tool, not a dictionary. The goal isn't to win debates. The goal is to make the next decision clearer.

Agile

What people say: "We need to be more Agile."

What it often means: "We're overwhelmed, priorities are unclear, and we want it to hurt less."

Reality check question: *What behavior needs to change in how we decide, plan, and respond under pressure?*

If it's real, you'll see: Clearer trade-offs, faster learning, fewer fire drills, more honesty.

Transformation

What people say: "We're doing an Agile transformation."

What it often means: "We're changing organizational charts, roles, and ceremonies and hoping behavior follows."

Reality check question: *What will leaders do differently when they get bad news?*

If it's real, you'll see: Fewer fear-based decisions, fewer hidden problems, more trust, more consistency.

Alignment

What people say: "We need alignment."

What it often means: "People disagree, but we don't want to talk about it out loud."

Reality check question: *What are we disagreeing about—goals, priorities, constraints, or trade-offs?*

If it's real, you'll see: Explicit decisions, fewer surprises, and less passive-aggressive behavior.

Empowerment

What people say: "We empower teams."

What it often means: "Teams are accountable, but leaders still control the real decisions."

Reality check question: *What decisions can teams make without permission, and what guardrails exist?*

If it's real, you'll see: Teams saying "no," "not yet," and "yes, but..." without punishment.

Accountability

What people say: "We need accountability."

What it often means: "Something went wrong, and we want someone to blame."

Reality check question: *Are we trying to learn or are we trying to punish?*

If it's real, you'll see: Clear ownership, clear expectations, and learning without scapegoats.

Velocity

What people say: "We need higher velocity."

What it often means: "We want more output, and we're hoping a metric will create it."

Reality check question: *What's slowing us down: unclear priorities, too much WIP, dependencies, rework, or decision delays?*

If it's real, you'll see: Smoother flow, less rework, and fewer last-minute scrambles, not just bigger numbers.

Value

What people say: "We need to deliver more value."

What it often means: "We're shipping activity, but customers aren't feeling it."

Reality check question: *What problem are we solving, for whom, and how will we know it helped?*

If it's real, you'll see: Customers noticing improvements, not just internal celebration.

Capacity

What people say: "We don't have capacity."

What it often means: "We're overloaded, but nobody wants to stop anything."

Reality check question: *What will we stop doing to make room?*

If it's real, you'll see: Fewer parallel initiatives, fewer meetings, and protected focus time.

Prioritization

What people say: "We need better prioritization."

What it often means: "Everything is urgent and no one wants to own the trade-offs."

Reality check question: *What are we willing to delay or drop, and who is making that call?*

If it's real, you'll see: Fewer "top priorities," clearer sequencing, and less thrash.

Operating Model

What people say: "We're moving to a new operating model."

What it often means: "We're reorganizing because it feels like progress."

Reality check question: *What behavior will change day to day: decisions, handoffs, feedback loops, conflict handling?*

If it's real, you'll see: Fewer bottlenecks and clearer ownership, not just new boxes on a slide.

Maturity

What people say: "We need to increase our Agile maturity."

What it often means: "We want compliance to a model because it's easier than dealing with behavior."

Reality check question: *What capability are we trying to build: learning speed, decision quality, flow, or well-being?*

If it's real, you'll see: Measurable improvements in outcomes and health, not higher scores on a rubric.

Best Practices

What people say: "This is best practice."

What it often means: "Someone important likes it, and we're skipping the thinking."

Reality check question: *Best for what context: constraints, goals, and what are we trading off?*

If it's real, you'll see: Adaptation, not copy-paste.

Scaling

What people say: "We need to scale Agile."

What it often means: "We want consistency, but we're about to standardize the wrong things."

Reality check question: *What must be consistent (goals, guardrails, language) vs. what must stay local (how teams work)?*

If it's real, you'll see: Shared intent and local autonomy, not bureaucracy.

Psychological Safety

What people say: "We value psychological safety."

What it often means: "We want people to speak up, but we don't want to hear bad news."

Reality check question: *What happens to the person who raises the hard truth?*

If it's real, you'll see: Earlier problem surfacing, less blame, and more honest conflict.

Continuous Improvement

What people say: "We're always improving."

What it often means: "We're always changing things, and nobody can tell if it helped."

Reality check question: *What did we change, what did we learn, and what's different because of it?*

If it's real, you'll see: Time-boxed experiments, reviews, and fewer repeated mistakes.

The Three Questions That Cut Through Almost Anything

If you only remember three questions from this appendix, make them these:

1. What problem are we solving specifically, for us, in our context?
2. What behavior has to change for this to work (especially from leaders)?
3. What will we stop doing to make room, and how will we know it helped?

If people can't answer those, you're not improving. You're performing. These are "clarity lines" you can use in the moment. They're designed to be direct without being combative.

For Leaders

- "Before we roll anything out, what behavior are we asking people to change, and what will I do differently first?"
- "What are we going to stop doing to make room for this?"
- "If this works, what will look different in 90 days that customers or teams will actually feel?"

For Coaches/Practitioners

- "What problem are we solving in *this* context, and what evidence will tell us it's improving?"
- "What's the smallest reversible experiment we can run instead of a full rollout?"
- "What's the one conversation we're avoiding that would make this real?"

For Team Members

- "What does that mean in practice? What will we do differently next week?"
- "What are we trading off if we say yes to this?"
- "I'm stuck/I'm behind/I need help, can we look at this together before it becomes a fire drill?"

Appendix B

RED FLAGS—ARE WE IMPROVING OR PERFORMING?

This appendix is a mirror. It's here to help you spot the difference between *real* improvement and performance that looks like improvement from a distance. Most organizations don't wake up and decide to waste time. They drift. They get busy. They get scared. They start optimizing for optics. These red flags aren't gotchas. They're signals. If you can see the signal early, you can correct it before it becomes culture.

The Big Pattern: Surface Change, Same Behavior

If you change labels, tools, roles, or ceremonies, but decisions, conflict habits, and leadership behavior stay the same—you're not improving. You're redecorating. If you remember one line from this appendix, make it this: *If the behavior didn't change, the system didn't change.*

Trend-Chasing Red Flags (The Whiplash Signals)

- "A new framework every 12–18 months."

What you'll see: Scrum → SAFe™ → Spotify → hybrid → "our own model."

What it usually means: Leadership is shopping for certainty instead of changing behavior.

Reality question: *What behavior will leaders change this time that they didn't change last time?*

- "Big launch, big noise, slow drift back."

What you'll see: Kickoff events, training, posters, new terms… then nothing sticks.

What it usually means: No protected behavior change, no reinforcement, no trade-offs.

Reality question: *What are we protecting for 30 days, and what will we stop doing to make room?*

- "The dip that never ends."

What you'll see: Productivity drops during the current change and never fully recovers before the next change hits.

What it usually means: Change saturation, constant relearning, and initiative overload.

Reality question: *What's the stop-doing list? What are we killing before we add anything?*

- "People joke about the change."

What you'll see: Eye-rolling, dark humor, quiet sarcasm.

What it usually means: Learned helplessness; people don't believe honesty is worth it.

Reality question: *What didn't work last time, and what would have to be true for this to help?*

Framework Performance Red Flags (Looks Like Agile, Isn't)

- "We're busy, but nothing finishes."

What you'll see: Lots of work started, little work done.

What it usually means: High WIP, too many dependencies, unclear priorities.

Reality question: *What can we stop starting so we can finish?*

- "Stand-ups are status theater."

What you'll see: "Still in progress" every day; no one asking for help; no coordination.

What it usually means: Low trust, fear of looking incompetent, or no one believes help is available.

Reality question: *What's blocked, what do you need, and who can help today?*

- "Retros are safe but useless."

What you'll see: "What went well/what didn't" with no behavior change after.

What it usually means: People are performing reflection without taking risks.

Reality question: *What will we do differently next week—and how will we know it happened?*

- "Metrics rotate every quarter."

What you'll see: New dashboards, new KPIs, new definitions of success.

What it usually means: Avoiding accountability for outcomes; chasing optics.

Reality question: *What are we measuring to learn, and what decision will this metric change?*

Leadership Red Flags (The Root Causes)

- "Bad news shows up late."

What you'll see: Surprises at the end; last-minute escalations; panic.

What it usually means: People don't feel safe telling the truth early.

Reality question: *What happens to the person who raises the hard truth?*

- "Everything rolls up to one person."

What you'll see: Constant escalations; leaders as bottlenecks; slow decisions.

What it usually means: Low trust, unclear guardrails, or leaders addicted to control.

Reality question: *What decisions can teams make without permission—and what guardrails do they need?*

- "Commitments without trade-offs."

What you'll see: Dates set by optimism; scope never moves; quality gets sacrificed quietly.

What it usually means: Leadership wants certainty more than reality.

Reality question: *If we keep the date, what are we trading off: scope, quality, or people?*

- "Hero leaders are celebrated."

What you'll see: Leaders swoop in, save the day, work late, and fix the crisis.

What it usually means: The system is fragile; heroics are compensating for structural failure.

Reality question: *What would we change so this never needs a hero again?*

Culture Red Flags (The Quiet Signals)

- "Polite agreement in public, cynicism in private."

What you'll see: Nodding in meetings, eye-rolling afterward.

What it usually means: Psychological safety is low; honesty feels pointless.

Reality question: *What are people not saying out loud, and why?*

- "People stop trying."

What you'll see: Minimal effort and numbness, "just tell me what you want."

What it usually means: Learned helplessness; repeated change with no payoff.

Reality question: *What would make it worth caring again? What friction can we remove this week?*

- "Top talent leaves first."

What you'll see: The improvement-minded people exit early.

What it usually means: Capability bleed; people don't see a future here.

Reality question: *Who are we losing, what do they have in common, and what are we doing that's pushing them out?*

The "Are We Actually Improving?" Reality Checks

- If you want a quick gut-check, ask these questions and listen to how hard it is to answer:
  - o What behavior changed in the last 90 days? (not what process changed)
  - o What did we stop doing to make room?
  - o What did we learn recently, and what changed because of it?
  - o Where does work get stuck, and what are we doing about it?
  - o Are customers feeling the difference?
  - o Are people going home with energy left?

If you can't answer these questions, you're probably performing improvement instead of doing it. When you spot a red flag, don't start a new initiative. Pick one behavior to change on purpose for 30 days. Make it visible. Review it honestly. Keep what helps. Drop what doesn't. That's how you move from graveyards to growth, without becoming the next trend.

Appendix C

BEHAVIOR-FIRST CORE BEHAVIORS (THE ONES THAT ACTUALLY CHANGE OUTCOMES)

Frameworks don't change outcomes. People do. This appendix is the behavior layer underneath everything in the book. These behaviors aren't inspirational posters. They're observable actions you can practice, reinforce, and protect, especially when pressure hits. Each behavior includes:

- What it looks like day to day.
- What it doesn't look like.
- How leaders accidentally kill it.
- How coaches/practitioners accidentally kill it.
- A 30-day practice sign
- Signals you can watch for (so it doesn't become vibes).

You don't need to do all of these at once. Pick one. Practice it for 30 days, review it, and repeat.

Tell the Truth, Even When It's Uncomfortable

What it looks like: Truth shows up early. Not as a dramatic confession at the end. It sounds like:

- "We're going to miss the date."
- "This isn't working."
- "I don't understand what success looks like."

- "I made a mistake."
- "I need help."

In a behavior-first environment, truth isn't punished. It's used. People respond to bad news with trade-offs and support, not humiliation.

What it doesn't look like:

- Pretending everything is fine until it explodes
- "Green Status" theatre (taking a green status signal as truth without confirming evidence)
- Sugar-coating reality so nobody gets upset
- Blaming the person who surfaced the issue
- "We'll figure it out later" as a strategy

How leaders accidentally kill it:

- Shooting the messenger (even once).
- Reacting to bad news with anger, sarcasm, or shame.
- Asking for honesty, then rewarding optimism.
- Using truth as ammunition in performance reviews.

If people only tell you the truth in private, you don't have honesty; you have fear management.

How coaches/practitioners accidentally kill it:

- Letting meetings stay vague to "keep things positive."
- Over-facilitating conflict instead of surfacing it.
- Hiding behind process ("let's take that offline") when the team needs clarity now.
- Treating honesty as a retro activity instead of a daily norm.

30-day practice sign: "Bad news early"

Pick one rule and protect it for 30 days:

- Bad news must be raised within 24 hours of noticing it.
- Every plan includes explicit trade-offs.
- Every status includes one risk and one ask.

Leaders: when bad news shows up, your first words matter. Practice: "Thanks for telling me early. What do you need? What trade-off are we making?"

Signals to watch:

- Risks surface earlier (not at the end)
- Fewer surprises
- Fewer fire drills
- More direct disagreement in meetings (less hallway eye-rolling)

Share Information Instead of Hoarding It

What it looks like: People share both explicit knowledge (docs, how-tos, data, checklists) and tacit knowledge (context, judgment, "here's what usually goes wrong," stories). Teams build collective intelligence. Nobody is a single point of failure by design.

What it doesn't look like:

- "Only Sam knows how that works"
- Knowledge as status
- Gatekeeping access to data
- "I'll do it myself, it's faster" as a lifestyle

How leaders accidentally kill it:

- Rewarding individual heroics over team capability.
- Promoting the "indispensable expert" who hoards knowledge.
- Creating fear around mistakes, which makes people hide what they know.
- Treating documentation as busywork instead of risk reduction.

How coaches/practitioners accidentally kill it:

- Pushing documentation without making it useful.
- Creating templates nobody reads.

- Focusing on tools (wikis) instead of habits (sharing in the flow of work).
- Ignoring tacit knowledge because it's harder to capture.

30-day practice sign: "Make the invisible visible"

Pick one rule and protect it for 30 days:

- Weekly "what-I-learned" share meetings (10 minutes, practical, no fluff).
- "If you touch it, you teach it" rule: every fix includes a short note for the next person.
- Rotate ownership of one critical area with pairing/shadowing.

Signals to watch:

- Fewer bottlenecks around one person
- Faster onboarding
- Fewer repeated mistakes
- More consistent decisions (less tribal knowledge)

Ask for Help Before Things Explode

What it looks like: Asking for help is normal. It's a sign of professionalism, not weakness. It sounds like:

- "I'm stuck."
- "I'm behind."
- "I don't know."
- "Can you sanity-check this?"
- "I need a second brain."

What it doesn't look like:

- Silent suffering until the deadline hits
- Stand-ups that are just "still-in-progress" theater

- People hiding uncertainty to look competent
- Last-minute heroics as the default recovery plan

How leaders accidentally kill it:

- Punishing uncertainty.
- Rewarding the person who "saves the day" instead of fixing the system.
- Creating a culture where asking for help is seen as incompetence.
- Overloading teams, so asking for help feels pointless.

How coaches/practitioners accidentally kill it:

- Treating stand-ups as status reports instead of coordination.
- Not teaching teams how to ask for help clearly.
- Letting "blocked" become a label instead of a conversation.
- Over-indexing on tools/boards instead of human signaling.

30-day practice sign: "Help early, help clearly"

Adopt a simple format:

- What I tried.
- What I'm blocked by.
- What I need.
- By when it matters

Example: "I tried X and Y. I'm blocked by Z. If we don't resolve it by Thursday, it impacts the release. Can we look at it for 15 minutes?"

Signals to watch:

- Fewer late surprises
- More pairing/collaboration
- Shorter cycle time for stuck work
- Less after-hours scrambling

Adapt Decisions When Reality Changes

What it looks like: Plans are not sacred. Reality is. People notice when assumptions break and adjust decisions without ego:

- "That's not true anymore."
- "The constraint moved."
- "We learned something, so the plan changes."
- "We're not doing that now."

Frameworks stay in the toolbox. They don't become an identity.

What it doesn't look like:

- Marching to the edge of a destroyed bridge because that's what the map says.
- Sticking to a plan to protect ego.
- Pretending the customer, market, or system didn't change.
- Copying case studies like they're recipes.

How leaders accidentally kill it:

- Punishing course correction as a lack of commitment.
- Demanding certainty instead of learning.
- Forcing teams to hit dates without trade-offs.
- Making it politically dangerous to change your mind.

How coaches/practitioners accidentally kill it:

- Treating the framework as the goal.
- Over-standardizing practices that should be local.
- Confusing consistency with compliance.
- Ignoring context because it's messy.

30-day practice sign: "Assumption check"

Once a week, ask:

- What assumption are we making that might be wrong?

- What changed in reality this week?
- What decision should change because of it?

Signals to watch:

- Fewer sunk-cost projects
- Faster response to customer feedback
- Fewer "we knew this months ago" moments
- More explicit trade-offs

Treat Well-being as a Design Constraint (Not a Perk)

What it looks like: Well-being is built into how work is designed:

- Sustainable pace
- Protected focus time
- Reasonable WIP
- Fewer fire drills
- Psychological safety as nonnegotiable

Pressure exists. But stress isn't normalized.

What it doesn't look like:

- "Always on" heroics
- Death-march weeks as normal
- "We care about well-being" while punishing "no"
- Treating burnout like a personal weakness

How leaders accidentally kill it:

- Adding work without removing anything
- Rewarding overtime and heroics
- Using urgency as a leadership style
- Ignoring team health signals until turnover hits

How coaches/practitioners accidentally kill it:

- Pushing "more ceremonies" into overloaded teams.
- Measuring success by adoption instead of sustainability.
- Ignoring workload and meeting load because it's "not Agile stuff."
- Treating well-being as outside the system.

30-day practice sign: "Reduce load before adding improvement"
Pick one rule and protect it for 30 days:

- Cap WIP
- Cancel one recurring meeting
- Protect two focus blocks per week
- Kill one low-value initiative

Signals to watch:

- Fewer mistakes and less rework
- Fewer people going quiet in meetings
- Fewer after-hours messages
- More stable delivery without panic

Lead as Service, Adaptation, and Transformation

What it looks like: Leadership is a behavior, not a title. Leaders:

- Remove obstacles
- Create clarity
- Model vulnerability
- Reward early truth
- Support local decision-making
- Grow people and evolve systems

What it doesn't look like:

- Command-and-control disguised as "alignment"
- Micromanagement
- Punishing dissent
- Treating people like resources instead of humans
- "Because I said so" decisions

How leaders accidentally kill it (yes, leaders can kill leadership):

- Making everything roll up to them.
- Hoarding information "for safety."
- Reacting defensively to feedback.
- Promoting sharp elbows over team-making.

How coaches/practitioners accidentally kill it:

- Trying to be the hero instead of enabling leaders to change.
- Protecting leaders from hard feedback.
- Doing leadership work for leaders (creating dependence).
- Focusing on process while ignoring power dynamics.

30-day practice sign: "One leadership behavior"
Pick one behavior and practice it for 30 days:

- Respond calmly to bad news.
- Ask trade-offs before commitments.
- Change your mind when evidence changes.
- Publicly reward someone who surfaces a problem early.

Signals to watch:

- Faster decisions with clearer intent
- Fewer escalations for routine issues
- More ownership and less fear
- Improved retention of your best people

How to Use This Appendix Without Turning It into Another Program

Don't roll this out. Don't make it a training. Don't turn it into a maturity model.

1. Pick one behavior.
2. Practice it for 30 days.
3. Review what changed.
4. Keep what helps.
5. Drop what doesn't.
6. Repeat.

That's *Stop Chasing Agile* in practice.

Appendix D

CONTINUOUS IMPROVEMENT THAT DOESN'T BREAK PEOPLE (A FIELD GUIDE)

Continuous improvement isn't a program. It's not a quarterly theme. It's not a poster that says "Kaizen" beside the coffee machine. It's a habit: learn, adjust, repeat, without turning your workplace into a permanent construction zone. Most organizations don't fail at improvement because they lack ideas. They fail because they overload people, chase novelty, and confuse activity with progress. Appendix D is here to help you improve in a way that actually sticks and doesn't burn everyone out in the process.

The Core Idea: Improvement Is Change with Memory

If you change things and don't learn, you're just thrashing. If you learn and don't change, you're just talking. Real improvement has three parts:

1. **A problem worth solving** (not a trend worth copying)
2. **A behavior change** (not just a tool change)
3. **A feedback loop** (so you can learn and adapt)

The Improvement Traps (and How to Avoid Them)

Trap 1: "We'll add this on top of everything else."

What it causes: Burnout, cynicism, and fake participation

Antidote: Improvement must displace something. If nothing stops, nothing improves.

Trap 2: "We need a big rollout."

What it causes: Theater, resistance, and a dip that never recovers

Antidote: Pilot, learn, adapt, and then scale what works

Trap 3: "We changed the process, so behavior will follow."

What it causes: Surface change, same habits, and same outcomes

Antidote: Name the behavior you're trying to change and protect it long enough to become normal.

Trap 4: "We measure output, so we must be improving."

What it causes: Speed without quality, delivery without value, and people without energy

Antidote: Balance outcomes, flow, and sustainability

The Three Lenses (Use These to Pick the Right Improvement)

Lens 1: Outcomes

Are customers feeling the difference? Are we solving real problems?

Lens 2: Flow

Is work moving from idea to done with less friction, less waiting, and less rework?

Lens 3: Sustainability

Is the system working without chewing people up?

If an "improvement" doesn't help at least one of these, be honest: it's probably noise.

What Improvement Looks Like When It's Healthy

1. Problems surface early: Not hidden until they explode. Not saved for a quarterly retro. Early.

2. Trade-offs are explicit: "We can do X, but then Y moves." No pretending.
3. Experiments are small and reversible: You don't need permission to learn. You need guardrails and honesty.
4. Leaders protect the space for change: If leaders keep injecting urgent work, improvement becomes a hobby. Then it dies.
5. People can say "this isn't working" without punishment: That's the difference between learning and performance.

The "Small, Local, Real" Approach (How to Improve Without the Circus)

This isn't a step-by-step recipe. It's a way to think. Start with one real pain. Pick something people feel every week. If it's not felt, it won't stick. Good starting points:

- Too many fire drills
- Work takes forever to finish
- Rework is constant
- Decisions are slow or unclear
- People are afraid to speak up
- Meetings eat the day
- Priorities change daily

Reality question: *What hurts the most right now, and what is it costing us?*

Make the behavior change explicit. Ask: *What do we need to do differently, as humans, for this to improve?*

Examples:

- "Ask for help within 24 hours of being stuck."
- "Name trade-offs before committing to dates."
- "Stop starting new work until we finish one thing."
- "Surface risks weekly, not at the end."

If you can't name the behavior, you're probably changing wrappers. Protect capacity (or don't bother). Improvement needs oxygen. Oxygen is time and attention. If you want improvement without burnout, you need:

- A stop-doing list
- Fewer parallel initiatives
- Fewer status meetings
- Less WIP

Leader reality question: *What will we stop doing to make room?*

Run experiments like a grown-up. An experiment is not "we tried it once." It's:

- Time-boxed
- Reviewed
- Kept, adapted, or killed

A simple experiment frame:

- **Hypothesis:** "If we do X, we expect Y to improve."
- **Time-box:** 2–6 weeks (long enough to learn, short enough to quit)
- **Measures:** 1–2 metrics + 1 human signal
- **Guardrails:** What we won't sacrifice (quality, safety, and well-being)

Review without blame. A review is not a trial. It's a learning loop. Ask:

- What happened?
- What did we learn?
- What do we keep?
- What do we change?
- What do we stop?

If the answer is "nothing," you didn't run an experiment. You ran a meeting.

What Leaders Must Do (Even If They Hate This Section)

If you have authority, your behavior is the multiplier. Healthy improvement requires leaders to:

- Reward early bad news
- Make trade-offs explicit
- Stop injecting "urgent" work without removing something
- Protect focus time
- Treat well-being as a constraint, not a perk
- Change their minds when evidence changes

If leaders don't change, improvement becomes theater.

What Coaches/Practitioners Must Do (So You Don't Become Part of the Problem)

Your job isn't to install a model. It's to help people learn. That means:

- Translating buzzwords into behavior
- Keeping experiments small and real
- Measuring impact in outcomes, flow, and well-being
- Building capability, not dependence
- Being willing to kill your own favorite idea

If you're more excited about the next framework than one better conversation this week, you're drifting.

What Team Members Can Do

You don't need permission to practice better behavior. You can:

- Ask for help earlier
- Tell the truth sooner

- Name trade-offs
- Reduce WIP locally
- Share knowledge
- Call out vague language with calm questions
- Protect your own boundaries and model sustainable pace

Small acts compound. That's how culture shifts.

The "Stop-Doing" List (The Most Underrated Improvement Tool)

If you want one practical move that changes everything, it's this: Make a stop-doing list. Some examples:

- Stop launching initiatives without killing old ones.
- Stop status meetings that don't produce decisions.
- Stop measuring success by busyness.
- Stop rewarding heroics that create fragile systems.
- Stop pretending dates are commitments without trade-offs.

Improvement is subtraction as much as addition. Continuous improvement is not about being perfect. It's about being honest, learning fast, and changing behavior on purpose, without breaking people. Change is constant. Panic is optional. Build a system that can learn. Protect the humans inside it. And keep taking the next small step.

Appendix E

METRICS THAT DON'T LIE (AND THE ONES THAT DO)

Metrics are tools. Not trophies. Not weapons. Not a way to pretend everything is fine. Used well, metrics help you see reality sooner, make better trade-offs, and learn faster. Used badly, they turn into performance art: people gaming numbers while the system stays broken. This appendix gives you a simple, balanced way to measure what matters, without turning your workplace into a spreadsheet cult.

First: The Rules (So Metrics Don't Become a Disease)

Rule 1: Measure to learn, not to punish: If people get punished for metrics, they will protect themselves. They'll hide problems, sandbag estimates, and game the system. You'll get prettier dashboards and worse outcomes.

Rule 2: No metric is the truth: Every metric is a flashlight. It shows part of the room. Use multiple flashlights.

Rule 3: Pair every metric with a question: A number without a question becomes a judgment. A number with a question becomes learning.

Rule 4: Trends beat snapshots: One week means nothing. Patterns over time tell you what's real.

Rule 5: Don't measure what you can't influence: If teams can't influence the number, it becomes demoralizing noise.

The Three Buckets That Matter

If you want behavior-first, well-being–focused agility, measure across three buckets:

1. Outcomes (for customers and the business)
2. Flow (how work moves from idea to done)
3. Well-being and sustainability (whether the system is breaking people)

If you only measure outcomes, you'll burn people out. If you only measure flow, you'll optimize delivery while missing value. If you only measure well-being, you might feel good while delivering nothing. Balance is the point.

Outcomes Metrics (Customer + Business Reality)

- Customer reliability (simple but brutal)

What it tells you: Do customers experience you as dependable?

Signals: Uptime, incident frequency, response time, service level agreement hits/misses, defect rates in production.

Good question: *Are we getting more reliable over time, or just louder about "delivery"?*

- Customer impact/adoption

What it tells you: Are people using what you ship?

Signals: Adoption rate, retention, usage frequency, support tickets by feature, churn reasons.

Good question: *What did we ship that customers actually noticed—and how do we know?*

- Time-to-value (from request to benefit)

What it tells you: How long customers wait for improvement.

Signals: Lead time from request → release → measurable benefit.

Good question: *Where does value get stuck—decision-making, handoffs, rework, approvals?*

Common trap: Measuring "features shipped" as value. Shipping isn't value. Outcomes are value.

Flow Metrics (The System Tells on Itself)

These are the most honest metrics in knowledge work because they expose bottlenecks and overload.

- Work in Progress (WIP)

What it tells you: How overloaded you are right now. Too much WIP = slower delivery, more context switching, more mistakes.

Good question: *What can we finish before we start something new?*

- Cycle time (start → done)

What it tells you: How long work takes once you begin.

Good question: *Why is this taking longer—dependencies, unclear requirements, rework, waiting?*

- Throughput (done per time period)

What it tells you: Delivery capacity over time.

Good question: *Are we getting more done because the system improved—or because people are sprinting toward burnout?*

- Flow efficiency (work time vs. waiting time)

What it tells you: How much time work spends stuck.

Good question: *Where does work wait, and who has the power to unblock it?*

- Rework rate

What it tells you: Quality and clarity problems.

Signals: Reopened tickets, bug fixes, "we had to redo it," repeated misunderstandings.

Good question: *What keeps causing rework, unclear decisions, rushed work, and missing feedback?*

Common trap: Using velocity as the main flow metric. Velocity is a local, gameable number. Cycle time and WIP are harder to fake.

Well-being and Sustainability Metrics (The Ones Everyone Pretends Aren't Real)

If you don't measure sustainability, you'll accidentally build a system that only works when people suffer.

- Fire drill frequency

What it tells you: How much chaos you're normalizing.

Good question: *How many "urgent" events did we have this month, and how many were preventable?*

- After-hours work/boundary violations

What it tells you: Whether a sustainable pace is real or just a poster.

Good question: *Is overtime the exception or the operating model?*

- Perceived workload (quick pulse)

What it tells you: Whether people feel overloaded.

How to measure: 1–5 pulse weekly/biweekly: "My workload is sustainable."

Good question: *What's making it heavy right now, and what can we remove?*

- Focus time/meeting load

What it tells you: Whether people can actually think.

Signals: Hours in meetings, blocks of uninterrupted time, context switching.

Good question: *Are we protecting focus, or stealing it with status theater?*

- Burnout risk signals

What it tells you: Early warning signs.

Signals: Irritability, silence in meetings, rising mistakes, absenteeism, turnover, and "numbness."

Good question: *What are we asking people to absorb on top of a full workload?*

Common trap: Treating well-being as "HR stuff." It's a system constraint. Ignore it, and the system fails.

Metrics That Lie (or Get You Lied To)

These aren't always useless. They're just easy to game and dangerous when used alone.

1. Vanity metrics
    - Story points completed
    - "% utilization" (often just a burnout target)
    - Number of initiatives launched
    - Number of trainings/certifications completed
    - Lines of code, tickets closed, hours logged

Why they lie: They reward activity, not outcomes. They push people toward looking busy. The worst metric of all: "Green status." If everything is green all the time, it usually means people are hiding reality.

Better question: *What's red that we're not saying out loud?*

2. A simple scorecard (one page, three audiences)

You don't need 40 metrics. You need a small set that you actually review and act on.

Team scorecard (weekly)

- WIP (current)

- Cycle time trend (last 4–8 weeks)
- Rework signal (up/down)
- Fire drills this week (count + cause)
- Workload pulse (1–5)

Team question: *What's the one change that would reduce friction next week?*

Coach/practitioner scorecard (biweekly/monthly)

- Are experiments running and reviewed? (yes/no + outcomes)
- Decision delay hotspots (where work waits)
- Psychological safety proxy: problems surfaced early? (examples)
- Capability growth: what did the team get better at?

Coach question: *What behavior changed, and what evidence shows it?*

Leader scorecard (monthly/quarterly)

- Customer reliability trend
- Time-to-value trend
- Fire drill trend
- Turnover/regretted loss
- Workload sustainability trend

Leader question: *What are we rewarding that's creating hidden damage?*

When the Dashboard Is Green, but Reality Feels Bad

This happens all the time. Here's what to do:

1. Ask for one concrete example of "things are fine."
2. Look at WIP and rework (they usually tell the truth).
3. Ask the team privately: "What are you not saying in the big meeting?"
4. Check fire drills and after-hours work.

5. Pick one behavior to change for 30 days and measure whether friction drops.

If the numbers look good but people look tired, your metrics are missing the cost. Metrics won't save you, but they can stop you from lying to yourself. Measure outcomes, flow, and sustainability. Use metrics to learn. Make trade-offs explicit. Protect people. And remember: if your system only works when people suffer, the system doesn't work.

Appendix F

CONVERSATION SCRIPTS (LEADERS/COACHES/TEAMS)

These are scripts for real life. Not "perfect words." Just language that helps you say the thing you're avoiding clearly, respectfully, and early. Use them as starting points. Make them sound like you. The goal isn't to win. The goal is to surface reality, make a better decision, and protect people from avoidable chaos.

1. "We're going to miss the date."

Best for: anyone (especially team members and team leads)

Say this: "Quick heads up: based on what I'm seeing, we're going to miss the date."

If you get pushback ("Just make it happen"): "I hear you. I'm not trying to be negative, I'm trying to be accurate early enough that we still have options. If we keep the date, we need to change something: scope, people, quality expectations, or sequence. Which trade-off do you want?"

If you want to keep it calm and factual: "Here are the top two risks and what I need to reduce them. If we can't change those, the date is a guess."

2. "What are we trading off?"

Best for: leaders, coaches, and anyone in planning

Say this: "Before we say yes, what are we trading off?"

If you get vague answers ("We'll figure it out"): "If we don't name the trade-off, we're not avoiding it, we're just pushing it onto the team as overtime and stress. What moves back if this moves forward?"

If someone insists everything is a priority: "Then nothing is. Pick the one you're willing to defend when the other work slips."

3. **"What will we stop doing to make room?"**

Best for: leaders and managers (but anyone can ask)

Say this: "What are we going to stop doing to make room for this?"

If you get "Nothing, we'll just add it": "Adding without removing is how we create burnout and fake deadlines. If nothing stops, this isn't a priority; it's a wish."

If you want to offer options: "Here are three candidates we could pause or drop. Which one hurts least?"

4. **"What does that mean in practice?"**

Best for: team members, coaches, and anyone hearing buzzwords

Say this: "What does that mean in practice? What will we do differently next week?"

If you get more buzzwords: "Totally. I'm still not clear on the behavior change. What will we start doing, stop doing, or decide differently?"

If someone gets defensive: "I'm not trying to poke holes. I'm trying to make sure we're not leaving this meeting with a slogan instead of a plan."

5. **"I need help."**

Best for: team members, leads, and anyone stuck

Say this: "I'm stuck and I need help."

If you want to be specific (recommended): "I've tried X and Y. I'm blocked by Z. If we don't resolve it by [time], it will impact [thing]. Can we look at it together for 15 minutes?"

If you're worried it'll look like failure: "I'd rather ask early than turn this into a last-minute fire drill."

6. **"This isn't working."**

Best for: anyone, especially in retros or 1:1s

Say this: "I don't think this is working."

If you want to make it safe and useful: "I'm not blaming anyone. I'm saying the current approach is creating [rework/delays/confusion/stress]. Can we look at what's driving it and try one small change for 30 days?"

If someone says, "That's just how it is here." "That might be true today. I'm asking if we want it to stay that way."

7. "We need to talk about how we handle bad news."

Best for: leaders, coaches, and team leads

Say this: "I want to set a standard: bad news comes early, and it doesn't get punished."

If you want to model the behavior: "If someone tells me we're behind, my job is to ask what they need and what trade-off we're making—not to shame them."

If you're rebuilding trust: "I know people may not believe this yet. I'm going to prove it by how I respond for the next few months."

8. "Let's kill this experiment."

Best for: coaches, leaders, and teams running improvements

Say this: "I think we should kill this experiment."

If you want to keep it learning-focused: "Not because anyone failed—because the results aren't worth the cost. What did we learn, and what are we keeping or changing because of it?"

If someone is attached to it (sunk cost): "We don't get points for sticking with a bad idea longer. The win is learning quickly."

9. "I disagree, and I want to do it out loud."

Best for: anyone (especially in low-safety environments)

Say this: "I disagree, and I want to say it out loud so we can make a better decision."

If you want to keep it respectful: "Here's what I'm seeing. Here's what I think we're missing. Here's the risk if we ignore it."

If you want to invite dialogue: "What am I not seeing from your side?"

10. "We're changing the wrapper, not the behavior."

Best for: coaches, practitioners, and leaders during rollouts

Say this: "I'm worried we're changing the wrapper, not the behavior."

If you want to make it concrete: "What decision will change? What conflict habit will change? What will leaders do differently when pressure hits?"

If you want to propose a next step: "Let's pick one behavior to protect for 30 days and measure whether friction goes down."

Mini-Scripts by Audience (Quick Access)

Leaders (three go-to lines)

- "What will we stop doing to make room?"
- "What behavior am I changing first?"
- "What will look different in 90 days if this worked?"

Coaches/practitioners (three go-to lines)

- "What's the smallest reversible experiment we can run?"
- "What's the one conversation we're avoiding?"
- "What changed in behavior, not just language, since last quarter?"

Team members (three go-to lines)

- "What does that mean in practice?"
- "What are we trading off?"
- "I'm stuck, I need help before this becomes a fire drill."

How to Use These Without Getting Fired (Low Psychological Safety Survival Guide)

If you're in a healthy environment, you can be direct, and it'll land. If you're in a low-safety environment, you can still tell the truth; you just need to be strategic about *how* you do it so you don't become the "problem person" for naming reality. This isn't about being fake. It's about getting results without becoming a casualty.

1. **Start with intent, not accusation**

Do this:

"I want to surface this early so we have options."

"I'm trying to protect delivery and reduce surprises."

"I'm not blaming anyone, I'm trying to make the trade-offs explicit."

Avoid this (even if it's true):

"This is stupid."

"Leadership doesn't get it."

"You're setting us up to fail."

Low-safety orgs punish tone more than content. Lead with intent so people can't easily label you as "negative."

2. **Use evidence and patterns, not vibes**

Do this:

"In the last three cycles, we slipped by 2–3 weeks each time."

"We have 14 items in progress and three people. That's why nothing finishes."

"We've had five fire drills this month. That's the cost."

Avoid this:

"I feel like we're always behind."

"This is chaotic."

Feelings matter, but evidence travels better, especially upward.

3. **Ask questions that force clarity (instead of making declarations); questions are harder to punish than statements.**

Try:

"What will we stop doing to make room?"

"What behavior is changing because of this?"

"What will look different in 90 days if it worked?"

"What are we trading off if we keep the date?"

If someone reacts badly, you can calmly say: "I'm asking because I want us to succeed. I'm not trying to challenge authority."

4. **Go private first when the room isn't safe**

If the culture punishes public disagreement, don't make your first stand in a town hall.

Do this:

- Raise it 1:1 with your manager first.
- Share a short written note (bullets, not a novel).
- Ask for permission to surface it: "Can I raise this in the meeting?"

This isn't cowardice. It's choosing the path that actually gets the issue addressed.

5. **Offer options, not just problems**

Low-safety environments often hear "problem" as "attack." Pair truth with choices.

Format:

"Here's what I'm seeing. Here are two options. Here's the trade-off."

Example:

"We can keep the date by cutting scope or accepting lower quality. If we keep scope and quality, the date moves. Which one do you want to own?"

6. **Don't die on every hill; pick the hill that matters**

If you challenge everything, people stop listening. Save your credibility for the moments that protect people, customers, or major risks.

Good hills:

- Unrealistic deadlines that create burnout

- Quality shortcuts that create customer harm
- Repeated fire drills that never get fixed
- Blame culture that hides problems

7. **Use "I" language, then anchor to impact**

Do this:

"I'm concerned we're setting a date without a trade-off."

"I'm noticing people are going quiet in meetings."

"I'm seeing rework increase."

Then anchor it: "That usually means problems surface late, and we pay for it in fire drills."

8. **Escalate with care (and protect yourself)**
   If you need to escalate, keep it boring:

   - Short summary
   - Evidence
   - Impact
   - Options
   - Request

Template:

"Here's the pattern. Here's the impact. Here are two options. I'm recommending option A. Can you confirm?"

And if you're in a truly toxic place: document key decisions and trade-offs in writing. Not to be petty, so reality can't be rewritten later.

9. **If you're the leader, prove safety through your reactions**

If you have authority, you set the weather. When someone brings bad news, your first move is everything:

- "Thanks for telling me early."
- "What do you need?"

- "What trade-off are we making?"
- "What do we learn and change so this doesn't repeat?"

If you punish truth once, you'll pay with silence for months.

10. The simplest rule

If you want honesty, make honesty survivable. That's it. That's the whole game.

BIBLIOGRAPHY

Argyris, C. (1990). *Overcoming Organizational Defenses: Facilitating Organization Learning*. Allyn & Bacon.

Argyris, C., & Schön, D. A. (1996). *Organizational Learning II: Theory, Method and Practice*. Addison-Wesley.

Argyris, C., & Schön, D. A. (1997). Organizational learning: A theory of action perspective. *Reis*, 77–78.

Bass, B. M. (1985). *Leadership and Performance Beyond Expectations*. New York Free Press.

Berriault, J. (2024). *Beyond the Framework: Cultivating Agile Growth*.

Bessen, J. (2015). *Learning by Doing: The Real Connection Between Innovation, Wages, and Wealth*. Yale University Press.

Brynjolfsson, E., & McAfee, A. (2014). *The Second Machine Age: Work, Progress, and Prosperity in a Time of Brilliant Technologies*. W. W. Norton.

Burns, J. M. (1978). *Leadership*. Harper & Row.

Deci, E. L., & Ryan, R. M. (1985). *Intrinsic Motivation and Self-Determination in Human Behavior*. Plenum.

Deming, W. E. (2018). *Out of the Crisis*. MIT Press.

Denning, S. (2018). *Why Agile Transformations Fail*. Forbes.

Derby, E., & Larsen, D. (2006). *Agile Retrospectives: Making Good Teams Great*. Pragmatic Bookshelf.

Doerr, J. (2018). *Measure What Matters*. Portfolio.

Dweck, C. S. (2006). *Mindset: The New Psychology of Success*. Random House.

Edmondson, A. C. (2018). *The Fearless Organization: Creating Psychological Safety in the Workplace for Learning, Innovation, and Growth*. Wiley.

Edmonson, A. (1999). Psychological safety and learning behavior in work teams. *Administrative Science Quarterly, 44*(2), 350–383.

Forsgren, N., Humble, J., & Kim, G. (2018). *Accelerate: The Science of Lean Software and DevOps.* IT Revolution Press.

Gibbons, P. (2019). *The Science of Organizational Change.* Phronesis Media

Gillis, A. (2008). *A Killing Art: The Untold History of Tae Kwon Do.* ECW Press.

Goleman, D. (2017). *Primal Leadership: Unleashing the Power of Emotional Intelligence.* Harvard Business Review Press.

Greeleaf, R. K. (1977). *Servant Leadership: A Journey into the Nature of Legitimate Power and Greatness.* Pualist Press.

Greenleaf, R. K. (1970). *The Servant as Leader.* Robert K. Greenleaf Publishing Center.

Grover, T. S., & Wenk, S. L. (2013). *Relentless.* Scriber.

Hackman, J. R. (2002). *Leading Teams: Setting the Stage for Great Performances.* Harvard Business School Press.

Halpern, C., Close, D., & Johnson, K. (1994). *Truth in Comedy: The Manual of Improvisation.* Meriwether.

Heifetz, R. (1994). *Leadership Without Easy Answers.* Belknap Press.

Heifetz, R. A., & Linsky, M. (2002). *Leadership on the Line: Staying Alive through the Dangers of Change.* Harvard Business School Press.

Herzberg, F. (1966). *Work and Nature of Man.* World Publishing.

Hiatt, J. (2006). *ADKAR: A Model for Change in Business, Government, and Our Community.* Prosci Learning Center.

Holyoak, K. J., & Morrison, R. G. (2005). *The Cambridge Handbook of Thinking and Reasoning.* Cambridge University Press.

Imai, M. (1986). *Kaizen: The Key to Japan's Competitive Success.* McGraw Hill.

Johensen, R. (2012). *Leaders Make the Future: Ten New Leadership Skills for an Uncertain World.* Berrett-Koehler.

Kerth, N. L. (2001). *Project Retrospectives: A Handbook for Team Reviews.* Dorset House.

Kotter, J. (1996). *Leading Change.* Harvard Business School Press.

Lencioni, P. M. (2002). *The Five Dysfunctions of a Team.* Jossey-Bass.

Maslach, C., & Leiter, M. P. (1997). *The Truth About Burnout: How Organizations Cause Personal Stress and What To Do About It.* Jossey-Bass/Wiley.

McGrefor, D. (1960). *The Human Side of Enterprise.* McGraw-Hill.

Norman, D. A. (2013). *The Design of Everyday Things.* Basic Books.

Northouse, P. G. (2021). *Leadership: Theory and Practice.* 9th ed. Sage.

Pfeffer, J., & Sutton, R. (2000). *The Knowing-Doing Gap: How Smart Companies Turn Knowledge into Action.* Harvard Business School Press.

Poppendieck, M., & Poppendieck, T. (2003). *Lean Software Development: An Agile Toolkit.* Addison-Wesley.

Ries, E. (2011). *The Lean Startup.* Crown Business.

Rost, J. C. (1991). *Leadership for the Twenty-First Century.* Preager.

Royce, W. W. (1970). Managing the development of large software systems. *Proceedings of IEEE WESCON, 26*, 328–388.

Schein, E. H. (1985). *Organizational Culture and Leadership.* Jossey-Bass.

Schein, E. H. (2013). *Humble Inquiry: The Gentle Art of Asking Instead of Telling.* Berrett-Koehler.

Seligman, M. E. (1975). *Helplessness: On Depression, Development, and Death.* W. H. Freeman.

Senge, P. (2006). *The Fifth Discipline: The Art and Practice of the Learning Organization.* Random House Books.

Stewart, M. (2009). *The Management Myth: Why the Experts Keep Getting It Wrong.* W. W. Norton.

Stokes, P. D. (2006). *Creativity from Constraints: The Psychology of Breakthrough.* Springer.

Taleuchi, H., & Nonaka, I. (1986). The new product development game. *Harvard Business Review, 64*, 137–146.

van Creveld, M. (1985). *Command in War.* Harvard University Press.

Willink, J., & Babin, E. (2015). *Extreme Ownership: How U.S. Navy SEALs Lead and Win.* St. Martin's Press.

Wooley, A. W., Chabris, C. F., Pentland, A., & Malone, T. W. (2010, October 29). Evidence for a collective intelligence factor in the performance of human groups. *Science*, 686–688.

INDEX

C

D

E

F